CW01652326

# From Hook to Plate: The State of Marine Fisheries

## A Commonwealth Perspective

Editors Richard Bourne and Mark Collins

Commonwealth Foundation

cpsu

CHEC

Published by the Commonwealth Foundation
Cover designed by WOW Creative
Printed by Hobbs the Printers, Totton, Hampshire, UK



The Commonwealth Fisheries Programme is being conducted over two years to provide civil society perspectives on the challenges facing marine fish stocks, and the livelihoods of coastal communities. Views and opinions expressed in this publication are the responsibility of the author(s) and should in no way be attributed to the institutions to which he/she is affiliated or the funders of this programme including the Commonwealth Foundation, DfID and AusAID.

The citation for this publication is: Bourne, R. and Collins, M. (eds) (2009). *From Hook to Plate: The State of Marine Fisheries. A Commonwealth Perspective*. London: Commonwealth Foundation. xii + 244 pp.

For further information contact:

Commonwealth Foundation
Marlborough House
Pall Mall
London SW1Y 5HY
United Kingdom

Tel: +44 (0) 20 7930 3783
Fax: +44 (0) 20 7839 8157
Email: geninfo@commonwealth.int
Web: www.commonwealthfoundation.com

British Library Cataloguing in Publication Data
A catalogue record for this book is available from the British Library

ISBN 978-0-903850-37-7

# Contents

# Message of support from Hon. Dr Abraham Iyambo, Minister of Fisheries and Marine Resources, Namibia

Namibia strongly supports efforts by the Commonwealth to minimise detrimental effects, such as IUU fishing, on fisheries through national, regional and international cooperation and collaboration with other players. The coastal area off Namibia is one of the most biologically productive marine environments in the world, supporting a rich marine life, including numerous fish, seabirds and marine mammals.

Namibia's primary commercial species include hake, monkfish, horse mackerel, pilchard, orange roughy, deep-sea red crab, rock lobster and Cape fur seals. About 95 per cent of the fish caught in Namibian waters is destined for exporting to Europe, Africa, Asia and the USA. Prior to Namibian Independence in 1990, uncontrolled fishing on a massive scale greatly reduced the abundance of all major fish stocks. During the 1960s, factory ships undertook fish processing at sea outside the 12 kilometers of jurisdiction of Namibia's fisheries administration, leading to the serious depletion of the pilchard and anchovy stocks. The late 1960s and 1970s saw the development of long distance freezer trawlers further increasing foreign interest in Namibia's offshore fishing grounds, and increasing reported hake catches from 50 000 MT in 1964 to 820, 000 MT in 1972.

Namibia is one of the countries in the world that has fully implemented its National Plan of Action on Illegal, Unreported and Unregulated fishing (NPOA-IUU). In addition, we made remarkable progress and played a prominent role in terms of our active participation in international and regional fisheries affairs and arrangements especially in the High Seas Task Force (HSTF), in SADC (Southern African Development Conference) marine affairs and governance issues and through our active involvement with Regional Fisheries Management Organizations (RFMOs) such as the International Commission for the Conservation of Atlantic Tuna (ICCAT), the Commission for the Conservation of Antarctic Marine Living Resources (CCAMLR), the South East Atlantic Fisheries Organization (SEAFO) and the Benguela Current Commission (BCC).

We will do our part to ensure that topical issues including high seas governance, climate change and its impact on fisheries and aquaculture, remain prominent on global fisheries agenda. There is no doubt that our active participation in regional and global fisheries affairs has contributed to the strengthening of Namibia's own fisheries management regime to be modern, responsive and relevant. It is therefore important to keep the momentum and consolidate our gains both at home and global level.

In terms of eradicating IUU fishing, the prevention of fish stock depletion and the promotion of sustainability, the Commonwealth may consider a few key initiatives. It can provide support to improve regional and inter-regional cooperation along with strengthening fisheries governance and legal frameworks to eliminate illegal fishing. It also could provide support to develop a regional monitoring, control and surveillance (MCS) strategy and a regional plan of action stressing the need to develop and adopt a National Plan of Action against IUU fishing. Also the Commonwealth Fund for Technical Cooperation (CFTC) may consider extending further assistance for capacity-building of small-scale fishers as well as promoting a network among academics, private sectors and non-governmental experts to supplement the work of governments in combating depletion of fish stocks, promoting sustainability strategies and the long-term viability of coastal vulnerable communities.

It is important that these ideas be promoted with key institutions such as the African Union and EU. A Commonwealth body should be established and should have teeth to identify and examine (having regard to Commonwealth as well as global initiatives) any further actions and measures that could be undertaken at the regional level to intensify the fight against IUU fishing, including the possibility of developing a regional agreement to strengthen cooperation in fisheries MCS and law enforcements.

# Message of support from Hon. Ezechiel Joseph, Minister for Agriculture, Lands, Fisheries and Forestry, St Lucia

The economy of St Lucia is based on the sustainable development of three major sectors with Agriculture (which includes Fisheries) as one of the main pillars.

The export of bananas to the UK market is the leading sub sector, however because of the many challenges this sub sector has been facing over the years, we have seen an increased interest in fisheries.

The significance of marine fisheries management in St Lucia is directly linked to the importance of marine fisheries to the socio-cultural and economic well being of St Lucians. In 2008, the fisheries sector employed roughly 2225 registered fishers, thus contributing to over three percent of the total employment in St Lucia. In many cases fishers are the main providers for their families and the industry also offers employment opportunities in other areas such as fish processing and vending; thus the total number of people dependant on fishing is much higher.

Therefore it is necessary to ensure that fish populations are harvested at their carrying capacity so as to secure the economic, social and nutritional benefits for future generations – as the term 'Sustainable Management' implies. This is part of the primary mission of the Department of Fisheries; therefore, the measures to achieve this are stipulated in the Plan to Manage Marine Fisheries of St Lucia (2006).

The Plan to Manage the Marine Fisheries outlines various actions to maintain the fish catch as high as possible without depleting the fish stock. These actions include regulating the amount of fishing activity on a temporal and spatial scale, regulating the number of fish removed from a fish stock by fishing and avoiding the removal of immature fish. However, one of the major principles of sustainable fisheries management is the implementation of an ecosystem approach for fisheries. Hence, there is great need to conserve and protect the habitats that fisheries depend on; these include mangrove wetlands, coral reefs and sea grass beds.

There are many challenges to overcome in order significantly to develop this sub sector. Some can be identified, such as improved technologies; reducing the average age of our fisher folks; the systematic introduction of value added products from our primary catch and the regulation of illegal, unreported and unregulated fishing.

While the Fisheries Development has developed programmes to be able to overcome these challenges, the lack of the correct technical and financial support is hampering its accomplishments.

Therefore there *must be* a sustainable livelihoods approach with new directions if we are to develop our Small Scale Fisher Folks.

## Message of support from Hon. Huw Irranca-Davies, Minister for Marine and Natural Environment, UK

I have great pleasure in commending this report to colleagues throughout the Commonwealth. Marine fisheries, and their fragile state, are of concern to all of us. With global warming affecting many species, and fish protein a key element in food security especially in developing countries, we cannot afford to delay collective action.

The United Kingdom is and will continue to be an active international partner in the effort to stamp out Illegal, Unreported and Unregulated fishing, and to promote sustainable management of deepwater and nearshore fisheries. It has an interest in supporting the trade in fish which is important for many developing countries, and in the long-term viability of traditional fishing communities.

We are delighted that the Commonwealth is recognising its own expertise and opportunity in this field. Its world-wide marine knowledge, and the danger its citizens face from overfishing and diminishing catches, put it in the forefront of research-based action. Individual countries can do much. The Commonwealth together, working through regional management bodies and in cooperation with the Food and Agriculture Organisation, can do more still, notably achieving better management of this valuable natural resource.

I hope this report will highlight these important issues to members of the Commonwealth, to identify opportunities and crucially to focus attention on the need for shared solutions. My own Government looks forward to working with Commonwealth partners at the Port of Spain summit, 2009, and beyond.

# Executive Summary

The marine capture fisheries of special concern to the Commonwealth comprise both deep-water and nearshore, including reef, fisheries. Because of the historical and geographical heritage of the Commonwealth its member states are responsible for many of the marine resources of the world. Nearly all are being depleted by industrial over-fishing, illegal fishing and a harmful race to fish a scarce resource. There is growing evidence of serious collateral damage to marine environments through physical damage to the sea bed, impacts on the trophic chain and climate change. This authoritative analysis is both a picture of what is happening, and a call to action by the Commonwealth – its leaders, its scientists, its fishers and its environmentalists – to save marine capture fisheries, and the many communities that depend on them for food and employment.

The picture is serious, but not hopeless. This book draws attention to successes in the struggle against illegal, unreported and unregulated fishing (IUU), to recovery in inshore stocks where marine protected areas (MPAs) are well-managed, and to philosophies and examples of rights-based management that engender a long term view of stewardship. In this field it sees a 21st century leadership role for the Commonwealth, including its many governmental and nongovernmental institutions.

Experts here describe the findings from a two year Commonwealth Fisheries Programme of which this book is a product. They include the impact of the World Trade Organization and its negotiations over fishing subsidies; the effect of the UN Convention on the Law of the Sea on fisheries management; the role of sustainability certification for consumers in protecting stocks; the significance of marine fisheries for protein intake and food security, especially in developing countries; management cooperation and the impact of traditional law on Pacific fisheries; the scale of IUU fishing, especially off the African coast, and the measures used to combat it; regional management in the Caribbean and the impact of climate change on Caribbean fisheries; the scope and value of marine protected areas (MPAs), promoted by the World Summit on Sustainable Development, 2002; the scale of depletion, both worldwide and within the waters of Commonwealth states; the case for a rights-based management strategy; and the potential role of the Commonwealth of Nations.

The editors and authors hope that this report, in demonstrating the wide-reaching importance and vulnerability of marine capture fisheries and the urgency of corrective measures, will stimulate the leaders of Commonwealth governments, who meet in Trinidad and Tobago in November 2009, to commit to further concerted and determined action.

## Acronyms used in the text

| | |
|---|---|
| ACP | African Caribbean and Pacific Countries |
| ACS | Association of Caribbean States |
| ATLAFCO | African Atlantic Fisheries Conference |
| AU | African Union |
| BINGO | Big International NGO |
| CA | Conservation Area |
| CARICOM | Caribbean Community |
| CBFM | Community-Based Fisheries Management |
| CFP | Commonwealth Fisheries Programme |
| CFP&R | Common Fisheries Policy and Regime |
| CFRAMP | CARICOM Fisheries Resource Assessment and Management Programme |
| CFTDI | Caribbean Fisheries Training and Development Institute |
| CHEC | Commonwealth Human Ecology Council |
| CHOGM | Commonwealth Heads of Government Meeting |
| CITES | Convention on International Trade in Endangered Species |
| CLME | Caribbean Sea Large Marine Ecosystem |
| COFI | Committee on Fisheries (of FAO) |
| COMAFAT | Comité Maritime de la Façade Atlantique |
| CPSU | Commonwealth Policy Studies Unit |
| CRFM | Caribbean Regional Fisheries Mechanism |
| CRISP | Coral Reef Initiative for the South Pacific |
| CTE | Committee on Trade and Environment |
| DFID | Department for International Development (UK) |
| DWFF | Distant Water Fishing Fleet |
| DWFN | Distant Water Fishing Nations |
| EBM | Ecosystem Based Management |
| EEZ | Exclusive Economic Zone |
| EJF | Environmental Justice Foundation |
| EU | European Union |
| FAO | Food and Agriculture Organisation |
| FDF | Food and Drink Federation (UK) |
| FFA | [Pacific Islands] Forum Fisheries Agency |
| GASSDD | Guidance for the Assessment of Small-Scale Data and Data-Deficient Fisheries |
| GBRMP | Great Barrier Reef Marine Park |
| GNP | Gross National Product |
| GPS | Global Positioning System |
| HIPC | Heavily Indebted Poor Countries |

| | |
|---|---|
| HMTC | Harmonised Minimum Terms and Conditions for Foreign Fishing Vessels |
| HSTF | High Seas Task Force |
| ICCAT | International Commission for the Conservation of Atlantic Tunas |
| ICTSD | International Center for Trade and Sustainable Development |
| IDPPE | Institutuo Nacional de Desentvolvimento de Pesca de Pequena Escala [National Institute for the Development of Small Scale Fisheries] |
| IUCN | International Union for Conservation of Nature and Natural Resources (World Conservation Union) |
| IOCARIBE | Intergovernmental Oceanographic Commission, Sub-Commission for the Caribbean and Adjacent Regions |
| IPOA-IUU | International Plan of Action on IUU |
| ITQ | Individual Transferable Quota |
| IUU | Illegal, Unreported and Unregulated [fishing] |
| JICA | Japanese International Cooperation Agency |
| LFDC | Low Income Food Deficient Countries |
| LMMA | Locally-Managed Marine Areas |
| LOSC | Law of the Sea Convention [more commonly UNCLOS q.v.] |
| LVFO | Lake Victoria Fisheries Organisation |
| MCS | Marine Conservation Society |
| MCS | Monitoring, Control and Surveillance |
| MDG | Millennium Development Goals |
| MFC | Maori Fisheries Commission |
| MPA | Marine Protected Area |
| MSC | Marine Stewardship Council |
| MSY | Maximum Sustainable Yield |
| NAFMAC | New Zealand Ministry of Agriculture and Fisheries and National Fisheries Management Advisory Committee |
| NEPAD | New Partnership for Africa's Development |
| NGO | Non-Governmental Organisation |
| NGR | Negotiating Group on Rules |
| nm | Nautical mile |
| NPOA-IUU | National Plan of Action on IUU |
| OECD | Organisation for Economic Co-operation and Development |
| OECS | Organisation of Eastern Caribbean States |
| OLDEPESCA | Latin American Organisation for Fisheries Development |
| OSPESCA | Organisation for Fisheries and Aquaculture of the Central American Isthmus |
| PICTS | Pacific Islands Countries and Territories |
| PIF | Pacific Islands Forum |
| PNA | Parties to Nauru Agreement |
| PNG | Papua New Guinea |

| | |
|---|---|
| PSA | Port State Agreement |
| QMS | Quota Management System |
| RFB | Regional Fisheries Body |
| RFMO | Regional Fisheries Management Organisations |
| RFS | Responsible Fishing Scheme |
| ROO | Rules of Origin |
| SADC | Southern African Development Conference |
| SEFSC | South East Fisheries Science Center |
| SIDS | Small Island Developing States |
| SIF | Stop Illegal Fishing Programme |
| SME | Small and Medium Sized Enterprises |
| SPC | Secretariat of the Pacific Community |
| SPS | Sanitary and Phytosanitary |
| TAC | Total Allowable Catch |
| TNG | Trade Negotiations Committee |
| TURF | Territorial Use Rights Fisheries |
| UNCED | United Nations Conference on Environment and Development |
| UNCLOS | United Nations Convention on the Law of the Sea |
| UNEP | United Nations Environment Programme |
| UNFSA | United Nations Fish Stocks Agreement |
| USMLT | United States Multi-lateral Treaty |
| VDS | Vessel Day Scheme |
| VMS | Vessel Monitoring System |
| WCPFC | Western and Central Pacific Fisheries Commission |
| WCPO | Western and Central Pacific Ocean |
| WECAFC | Western Central Atlantic Fisheries Commission |
| WTO | World Trade Organization |
| WWF | World Wide Fund for Nature |

# 1 Introduction

Richard Bourne and Mark Collins

The state of marine capture fisheries, of concern to the 47 Commonwealth member states which have shores facing the high seas, is extremely fragile. As described by Rashid Sumaila and William Cheung in one of the chapters of this book, industrial overfishing has led to a world-wide crisis in the oceans. There is also a crisis of depletion which affects the nearshore subsistence fishers, whose contribution to the informal economy, nutrition and community cohesion along the coasts is often overlooked by economic analysts.

The chapters which follow pick out some of the key issues, and point to ways in which they may be resolved. They are one of the products of a two year Commonwealth Fisheries Programme, undertaken by three Commonwealth agencies with funding support from the United Kingdom and Australian governments on behalf of all members.

This report has been prepared in time for a summit meeting of Commonwealth leaders in Trinidad and Tobago in late November, 2009 but will have a more lasting value. It is a state-of-the-art account of issues which have particular resonance for Commonwealth countries, most of which are islands and where only six are landlocked. It will be backed by a call for Commonwealth leadership to take forward action by the association itself, in concert with the Committee on Fisheries of the Food and Agriculture Organisation, by support for regional fisheries management, by stopping Illegal, Unreported and Unregulated (IUU) fishing and by action within individual countries.

Although the Commonwealth is only a segment of the world community, its maritime history means that it is peculiarly well-placed for leadership in marine affairs, and its Exclusive Economic Zones (EEZs) cover millions of square nautical miles. Its fleets, the experience of its marine scientists and the record of its governments – especially in promoting the UN Convention on the Law of the Sea (1982) and in the High Seas Task Force (2003) – are testament to its capacity and expertise. For some years the Commonwealth Secretariat, which services governments, has been providing technical assistance and small-scale funding to governments and regional organisations.

The chapters which follow start by describing the findings from the Commonwealth Fisheries Programme, conducted by the Commonwealth Foundation, Commonwealth Human Ecology Council (CHEC) and Commonwealth Policy Studies Unit (CPSU) over two years. The Commonwealth Foundation is the intergovernmental body which promotes civil society. The Commonwealth Human Ecology Council is the Commonwealth non-governmental organisation concerned with the environment. The Commonwealth Policy Studies Unit is a think-tank which forms part of London University's Institute of Commonwealth Studies.

The chapters have been contributed by experts in their fields, and cover: findings from the programme; WTO negotiations on fishery subsidies; the Law of the Sea; the IUU issues;

regional management in the Caribbean; cooperative approaches in the Pacific; New Zealand as a case study in the evolution of national governance; fisheries and food security; fish supply chain issues for developing countries; customary law and inshore fisheries in Pacific Island states; the impact of climate change in the Commonwealth Caribbean; Marine Protected Areas as a strategy for sustainability; the vulnerability and sustainability of fisheries by marine regions around the world; the case for a rights-based strategy; and potential roles for the Commonwealth.

This fisheries programme originated in discussions in London in early 2007, was developed in June of that year and was implemented in 2008–09. The parties concerned believe that the Commonwealth, as a facilitating body with convening power in the world community and specific expertise, can go beyond its important role in small-scale technical assistance. To help resolve the challenge of marine capture fisheries it could bring together marine scientists, fishers and governments, working at various levels, nationally, regionally and globally. In the still relevant words of Sir Shridath Ramphal, Commonwealth Secretary-General from 1975 to 1990, "The Commonwealth cannot negotiate for the world, but it can help the world to negotiate". Each partner has had a different role in delivering the programme, whose outputs are publicly available on a website at *www.commonwealthfisheries.org*.

Those promoting the Commonwealth Fisheries Programme hope that this report will be widely read, and that it will help to galvanise the Commonwealth and its various actors to play a major part in the conservation and sustainable management of our marine capture fisheries. The Commonwealth has much to offer in resolving this global crisis.

# 2 Establishing the Context for Enhanced Commonwealth Support of Small Fisheries*

Nicholas Watts and Richard Bourne

## Introduction

The Commonwealth Fisheries Programme,[1] which formally began in the middle of 2008, comprised three *case studies* undertaken by the Commonwealth Human Ecology Council (CHEC); three *study tours*, undertaken by the Commonwealth Policy Studies Unit (CPSU); and certain civil society consultations supported by the Commonwealth Foundation. This chapter attempts to draw out key issues from the case studies and study tours of significance for Commonwealth fisheries policy.

The CFP case studies, undertaken in 2008–9 in Belize, Fiji and Sierra Leone, investigate the sustainable management of small fisheries[2] with a focus on the contribution of rights-based co-management to fisherfolk livelihoods. The analyses look at the multi-level governance of fisheries from the bottom-up, using intensive, structured conversations with fisherfolk and key stakeholders in local fishing communities, as well as interviews with key national and regional experts. The purpose is to identify those best practices in the governance of coastal, artisanal fisheries that are arguably transferable to other Commonwealth countries and beyond.

The selected countries have high potential for transferable lessons. Belize and Fiji have Equator Initiative Awards for their management of marine reserves from the 2002 Johannesburg World Summit for Sustainable Development, using distinct approaches. The Belize marine reserves are good examples of management delegation to NGOs, in consultation with local stakeholders, and raise generic governance questions – about fisher participation, the role of national and international NGOs, and national policy capacity development. Fiji, a country highly dependent on fisheries (Veitayaki, 1995), won the Equator Award for its community-based system of Local Marine Managed Areas (LMMAs), which were to be the focal point of the case study, and when Fiji was chosen for as a case study there was a promise, not kept, of democratic elections by March 2009. In Sierra Leone the significance of a sustainable fisheries policy was tied to the prospect that it might contribute about 5 per cent to domestic GNP. By conservative estimates the fish stocks are worth US$735m and

*Both the case studies and the study tours have enjoyed privileged access to the policy communities in the countries included in the Commonwealth Fisheries Programme, for which we are immensely grateful. We would also like to mention in particular the help and advice of Sandra Grant and James Azueta in Belize, Aisie Lucan, Raymond and Reynold Johnson in Sierra Leone, and Joeli Veitayaki and Bill Aalbersberg in Fiji on the case studies and Marta Lang and Geoffrey Wandesforde-Smith for their comments. For full acknowledgements, please see the country reports on the programme website at *www.commonwealthfisheries.org*

could yield US$59m in annual income (DFID *et al.*, 2007). The stocks are at risk, however, from Illegal, Unreported and Unregulated (IUU) fishing, both by outside pirate fishers and by local artisans forced to catch juveniles by the ravages of illegal trawlers (EJF, 2009).

To assess the capacity for sustainable governance of small-scale artisanal fisheries and the coastal communities that depend on them the *case study* fieldwork combines expert interviews, meetings with fishers (both community meetings and focus groups) and, where appropriate, participatory mapping of the fishery to identify changing patterns of effort and catch over time. The intent is to portray the experience of the fisherfolk in their own terms, and to explain how this interacts with national and regional governance mechanisms. Governance is enhanced if these extra-local mechanisms are seen as effective and legitimate but impeded if senior policymakers fail to accommodate and respond to fisherfolk concerns.

The study tours examine innovative approaches to fisheries management that are significant because they optimise returns and sustainability. In three countries with significant marine fisheries in each of three important Commonwealth regions – East Africa, the South Pacific, and the Caribbean – visitors took advice from governments, scientists, fishers and environmental bodies. The tours involved eight or nine persons in each region, drawn from environmental NGOs, fishers, academics, and fishery agencies. There were fortnight-long tours to Tanzania, Mozambique and Mauritius (2–14 November 2008), to Papua New Guinea, Vanuatu and Samoa (16–29 November 2008), and to St Lucia, Trinidad and Tobago, and Belize (24 January–4 February 2009).

The study tours provide snapshots of salient issues related to marine fisheries and compare local approaches between neighbouring countries. The three different regional teams reached collective conclusions for the CFP at the end of each tour, synthesising group discussions held at the end of each country visit. The tours also contributed to capacity building among team members.

## Governance at Local and National Level, Including the Involvement of Fisherfolk and Communities

The case studies show that Belize and Fiji are building on their successful track records with marine protected areas, but with some significant variations, and the occasional setback.

In Belize an extensive range of marine protected areas is mostly co-managed by delegation to local NGOs in partnership with either the Fisheries or Forests Department (Fairweather-Morrison, 2006). The MPAs have been successfully established. Since the 2002 Equator Awards, however, problems of monitoring and enforcement have persisted, the latter in large part, according to local fishers, because of incursions by IUU fishers from Guatemala and Honduras. Also, the fishers claim they have inadequate access to decision-making processes in the MPAs.

In Fiji rapid expansion of the community-based programme of Locally-Managed Marine Areas (LMMAs) stems from their success, measured both in terms of increased harvest and by the engagement of local communities in monitoring and analysis of the programme. Their networking across the country and region has also led to take-up elsewhere. The programme is also an example of supportive and responsive but not *dirigiste* involvement by the regional University of the South Pacific in Suva. The LMMAs are based on the traditional system of *qoliqolis*, which makes a coastal fishing area part of a community's customary property rights. Creation of an LMMA requires a request from the community, which helps ensure a transparent, bottom-up approach. Community implementation of the programme, with support from the University and networks of NGOs, is also relatively cheap. The Pacific *study tour* underlined the *case study* finding that, where government resources are low and communities widespread, community-built solutions are critical. Success depends on utilising and supporting traditional village governance systems so that, once an LMMA is requested, effective facilitation and follow-up generate the ownership and momentum that are critical to community design, implementation, *and* enforcement of measures to limit take and encourage recovery.

The Fijian clam fishery illustrates the success of the LMMA model. With the fishery in decline, several managed areas (*tabu*) were set up, having at their core community workshops to bring together local fishers, Fisheries Department people, and experts in participation and Fijian Affairs. With support from USAID and the MacArthur and Packard Foundations, the programme expanded from one site to more than 225 in Fiji and 450 in the western Pacific. A mid-term review of this programme recommended community-based monitoring and this policy was supplemented by a government initiative to support ethnic Fijian fishers (Tawaki *et al.*, 2001). The overall results are a 250–300 per cent increase in productivity in the overflow take area and a 10 per cent rise in household incomes. These are impressive indicators of the effectiveness of community-based adaptive management, where traditional knowledge and rapid assessment are combined. The key is community recognition of the need for action followed up by the training of fishers to monitor a fishery, so that they in effect become 'fisheries scientists' themselves.

It is deeply ironic that the democratically-elected government of Fiji was in the process of according property rights in the LMMAs to the traditional communities that had exercised them historically when a *coup d'etat* occurred in late 2006. The allocation of these property rights was given as one pretext for the coup. The LMMAs remain, however, and in the language of one interviewee are 'resilient governance mechanisms in a context of uncertainty'.

In Sierra Leone, by contrast, there is a dearth of policy tools, instruments and institutions to support a co-management approach to small fisheries, which have been reviving since the end of civil conflict in 2001, in the face of serious challenges (Thorpe *et al.*, 2008). A new fisheries policy in 2003 allows for open access, but with no specific rights for established fishing communities.

There is, however, consultation with local fishing communities in Sierra Leone in policy development, and fishers have recently been organised into trade unions; one (SLAAFU) focuses on the northern coast, the other (SLAFU) on the southern coast. The unions take an active role in conflict resolution, helping to explain government fisheries policy, its rules and regulations, to their members. This is important, because it involves issues like the control of mesh sizes and fishing methods, closed seasons, and the introduction of Marine Protected Areas (MPAs).

Sierra Leone suffers from limited capacity in the ministry responsible for fisheries, and it has inadequate information for planning. With over 5000 boats, and four enumerators, the quality of baseline catch data is questionable. This in turn makes problematic any effective evaluation of its measures to enhance the management of the country's fisheries.

The study tours show variations in the qualities of and capacities for governance among the nine different countries visited, with the key variables being fisheries legislation, the numbers employed in relevant Ministries and Fisheries Divisions, levels of investment in community-based management by NGOs, and the degree of cohesion of fishers and local communities.

In Trinidad and Tobago, for example, the fisheries legislation dating from 1916 does not anticipate the need to manage an open access fishery in which some species, notably kingfish, snapper and yellow fin tuna, are over-exploited. Although the populations of St Lucia (approx 160,000) and Vanuatu (approx 215,000) are not that dissimilar in size, the staffing of government fisheries agencies is markedly different – 35 in St Lucia and less than a third of that in Vanuatu, most of whom are donor-funded. This means that Vanuatu is highly dependent on chiefs and community organisations for the management of inshore fisheries. This risks overburdening voluntary groups, there, although in Samoa the strength of the traditional chiefly culture and strong government and international donor leadership has underpinned the successful development of fishery reserves and no-take zones. Mozambique illustrates the value of government technical support for small-scale fishers, with advice on making low cost nets, the use of lead weights from car batteries, and life jackets created from empty plastic bottles.

In general, apart from Belize, fishers themselves were unable to have any significant input to national fisheries management in the nine countries visited on the study tours. The reasons for this include the fragmentation and small size of the fishers' organisations (as seen in Mozambique); the numerical and economic weakness of fishers compared with other socioeconomic interests (as in Trinidad and Tobago, where there is a diverse economy and powerful energy sector); and poorer educational levels than national averages.

## Regional and International Management

The case studies show variation in the quality of fisheries management by region, chiefly influenced by the extent to which Commonwealth states are integrated with their neighbours. Regional integration is nowhere well-established,[3] but Belize, for example, is an effective partner in both Caribbean and Central American fisheries management systems, and this despite conflicts in Belizean waters with both Guatemalan and Honduran fishers. In Africa, although Sierra Leone shares common problems with its non-Commonwealth neighbours, Guinea, Senegal and Liberia, and hosts a large community of migrant fishers from Ghana regional integration is limited. The government belongs to a subregional fisheries commission based in Senegal, as well as to ATLAFCO (The African Atlantic Fisheries Conference) and COMAFAT (Le Comité Maritime de la Façade Atlantique), but these are not sufficient to cope with overlapping national jurisdictions. In the Pacific, Fiji, in its democratic phases, participated in the fisheries work of the Pacific Islands Forum (PIF), particularly on regulation of the tuna fishery through the Pacific Islands Forum Fisheries Agency. In 2007, the PIF endorsed 'the development of national tuna industries, in the context of a phased introduction of rights-based management arrangements supported by an appropriate management and regulatory framework' (Pacific Islands Forum Secretariat, 2007). It also backed the conservation and management policies of the Western and Central Pacific Fisheries Commission (WCPFC), and the negotiations for a 'best-practice South Pacific Regional Fisheries Management Organisation'.

The study tours highlight several issues beyond weak cooperation between neighbouring states, which clearly affects fisheries enforcement in St Lucia and St Vincent and the development of a joint sustainable fisheries strategy for Mozambique and Tanzania. First, political divisions in some regional bodies reduce their effectiveness. CARICOM and the Caribbean Regional Fisheries Management Organisation, for example, have been plagued by quarrels over a common fisheries zone, because some states wanted to retain exclusive national fishing areas. Second, there are overlapping regional jurisdictions, as is the case off the East African coast, where the Southern African Development Community (SADC), the European Union and five Francophone states (Mauritius, Réunion, Comoros, Madagascar and Seychelles) have overlapping interests. Third, many regional institutions are underfunded and understaffed, and lack regional champions. The advantage of regional champions can be seen in the leadership Namibia has shown within SADC, enabling the hosting of a SADC conference on IUU in Windhoek in mid-2008 and the consequent facilitation by CFP personnel of a civil society statement. At international level, the Committee on Fisheries of the FAO ought also to provide leadership on marine fisheries but is hampered by the inability of smaller and poorer developing countries to take part. For example, at the March 2009 meeting of the Committee in Rome, there was no representative from five of the nine countries visited on the study tours, or from any of the case study countries.

## Funding for Fisheries, for Recovery and for Strategies of Sustainability

Funding is intimately, but sometimes perversely, related to capacity. The case studies show the benefits derived in Belize and Fiji from the international significance of their biodiverse rich coral reef ecosystems. International NGOs with substantial funding and expertise play a major role in both countries, and both have enjoyed substantial funding from inter-governmental sources. This raises, however, challenges of dependency and discontinuity. Governments over-reliant on NGOs for governance and management functions fail to develop sufficient capacity of their own. And, if externally supported mechanisms of marine governance, such as the Coastal Zone Management Agency and Institute in Belize turn out to be time-limited projects, the institutional infrastructure for supporting small-scale fisheries and coastal biodiversity is subject to sudden rupture. In Sierra Leone, by contrast, where large-scale donor projects have not proved sustainable in the past, the present engagement by the African Development Bank in developing state-of-the-art landing sites with appropriate health and safety facilities looks more promising. Here, the big international NGOs are less prominent, although the Environmental Justice Foundation has mounted an effective, and locally celebrated, campaign to combat IUU fishing (see below).

The study tours spotlight different funding issues. They show, for example, that some governments are making significant incomes from tuna licence fees, estimated to be worth US$60m a year in Papua New Guinea and up to US$1.8m in Vanuatu, although this income is rarely ploughed back into fishery management or sustainability. It simply contributes to general Treasury revenue. There is also a danger that attractive incomes will encourage over-fishing to supply new processing plants, like those at Madang in Papua New Guinea, or near Port Vila in Vanuatu, built without much apparent regard for stocks. The costs of fisheries agencies in smaller countries may be carried in part by donors, as when Peace Corps volunteers are attached to the Fisheries Division in Vanuatu. In Mauritius and some Caribbean states fisheries divisions have been supported by JICA, the Japanese aid agency, support which can have strings attached. In Pacific island states and in Belize MPAs funded by international NGOs are agents of sustainability. But in Samoa village committees bear the costs of management, through a Trust Fund for their own contributions to MPAs and those of CRISP, an NGO concerned for coral reefs in the Pacific. Although MPA managers, thus, benefit from interest from the fund, NGO support can also be erratic, or short-term, and may give priority to external rather than national interests.

The Fiji case study, for instance, shows that big international NGOs (or 'BINGOs') wiped out or overtook the Fiji-based Fiji Environment Group. In Vanuatu the Reef Check NGO, which aims to train local people to check the health of a reef and its creatures, cannot finance work beyond this initial training. The ideal of fully self-sufficient MPAs is more nearly approached at Nguna Pele, in Vanuatu, where a sustainable and ongoing local income comes from providing a low cost boat service between neighbouring islands. In St Lucia, the Soufriere Marine Management Association is self-sufficient thanks to incomes from fishers, yachts, tourism and hotel interests.

## Illegal, Unreported and Unregulated (IUU) Fishing

The case studies affirm that IUU fishing remains a problem on the high seas and inside EEZs, where there is significant poaching by illegal foreign trawlers of tuna and high value stocks. It is also a problem with near shore and reef species. In Fiji, for example, the University of the South Pacific and the World Conservation Union (IUCN) found that the Fisheries Department did not use its own rules to prevent poaching. In Belize, where many fishers have dual nationality and part of the territory is subject to a land claim from Guatemala, Belizean fishers have a long history of protest against nearshore piracy by Guatemalan and Honduran fishers. Sierra Leone suffers from serious IUU activity by South Korean trawlers in particular, and meetings with local fishers in the communities of Bonthe, Tombo and Goderich show strong local support for banning IUU trawlers from national waters. The country's most conspicuous recent success in combating IUU fishing occurred when the Environmental Justice Foundation helped the Sierra Leone Navy track down and arrest an illegal trawler, filming the fishers boxing up Sierra Leone fish as 'Korean' for delivery to ports in the European Union. The SLAFU union also exerts pressure on the government to crack down on local fishers who are working illegally and helped secure the arrest of 84 boats, but faced violent acts of revenge against union officials when crews were released by the government. Tensions were exacerbated when the Minister also announced that, from 1 June 2009, mesh size regulations would be strictly enforced, leaving the union to find replacement nets for 85 per cent of its members, who did not have the resources to upgrade. The options were to cease fishing, or to fish illegally, and less safely, at night. The alternative fear was that, if the government rescinded the regulations, the result would be a 'rush to fish'. The larger lesson is that in the fragile post-conflict state of Sierra Leone any conflict within the fishing community has a serious potential for escalation.

Many countries are now installing Vessel Monitoring Systems (VMS). In theory, these allow government monitors to regulate vessels within EEZ waters and prevent deep water IUU fishing. In practice, however, the scale of the problem is so vast that with limited aircraft surveillance and patrol boat availability there are relatively few checks on catches. The real scope of the problem is, therefore, hard to estimate. In Tanzania, the claimed catch of 15,000 tons by non-Tanzanian fleets within the EEZ is thought to be a considerable underestimate. The Mozambican government estimates that there are some 200 foreign vessels regularly working in the 500,000 square miles of its EEZ and that its IUU losses are US$200m a year, mostly in tuna. In Mauritius, with 1.9m square miles of EEZ and only one plane to carry out inspections three times a week intelligence is plainly limited. Enforcement is not entirely absent, however. In the Pacific, recently, countries have been watching with interest Tongan demands that US$2.5m in compensation be paid for a Taiwanese vessel illegally fishing in Tonga's EEZ when caught by a New Zealand ship.

## Issues Affecting Small-scale and Artisanal Fishing

The focus in the Fiji *case study* on LMMAs shows fishers acknowledging that they engage in some proscribed activities. The same observation applies to Sierra Leone. In both cases their intent is to stop this activity, based on understanding its sustainability effects, and this was coupled in Sierra Leone to requests for help obtaining nets of proper mesh size.

The study tours also make it clear that artisanal fishers link their perceptions of the viability of their livelihoods to other issues. In the Pacific, for example, where nearshore fisheries are overexploited and large fish commercially extinct, enhancing nearshore fisheries is tied to poverty alleviation. The Fiji government, committed to protecting 30 per cent of its inshore fishing areas, is spending US$4000 monitoring each LMMA site in a current project. The popularity of the LMMA approach rests on the convergence of poverty alleviation with sustainable fishing (Leisher, van Beukering and Scherl, 2007). A comparison of households involved in LMMAs with others not involved, shows a doubling of relative income in the former. Data from the Fiji case study underline the willingness of communities to take a long view. Nearly 80 per cent of those interviewed took the view that they wanted to 'have fish for [their] kids and grand-kids', showing that issues of culture, identity and community are as important as pure economics or the need for food security, and in some cases more so.[4] Similar sentiments can be found in the comments Samoan villagers made to Pacific study tour participants.

The case studies and study tours also support the view that, while small-scale and artisanal fishers are important in many coastal communities, they suffer from lack of government recognition, which translates into fewer practical investments in refrigeration, modern landing sites and transport to markets. Mozambique is exceptional in this regard, because the government-backed Instituto Nacional de Desenvolvimento de Pesca de Pequena Escala (IDPPE, or National Institute for the Development of Small-Scale Fishing) carries out research, offers advice to fishers and fishing associations, and promotes the interests of some 140,000 small-scale fishers working on the coast and inland waters. Exploitation of artisanal fishers by middlemen is an issue in the Caribbean and Indian Ocean. While many governments are aware of the royalties they can earn from tuna licences, few have an accurate picture of the food and employment value of coastal subsistence fishers.

The opportunities for artisanal fishers to get fish to more distant populations, let alone for export, are inhibited by the lack of modern equipment and – for export to countries in the European Union – by the difficulty of meeting stringent hygiene requirements. Thus, it was important for fisheries policy in Sierra Leone, for example, that EU health requirements were satisfied so its fish could be exported to Europe. Other handicaps stem from the lack of resource, organisation and educational levels among the fishers themselves. Where fishers have got together to establish cooperatives with quality processing equipment – as in Belize where two major cooperatives export lobster to North America – the fishers themselves are earning respectable rewards. Low-interest loans to fishers' organisations can make a difference. Countries as different as Mozambique and Samoa have shown support for small-scale fishers, and their need for eco-sustainability. Mauritius has provided financial aid for fishers when storms prevent them from pursuing their occupation. Both Belize, where fishers sometimes act as guides to tourist boats, and Fiji, where there is a significant partnership between one resort and a marine protected area, are among the countries trying to reconcile the interests of tourists and fishermen. This can provide additional income for fishing communities, though sometimes fishers will complain that tourist interests are driving a conservationist MPA strategy to the detriment of small-scale fishers' interests.[5]

## Capacity Building

Capacity building has long been recognized as an essential precondition for improvement in resource management in developing countries, and in the particular case of small fisheries there is agreement about this in all sectors: fishers, NGOs, marine scientists, government fishery divisions and regulators. It involves both infrastructure and training.

The African Development Bank, for example, is building infrastructure which will enable Sierra Leone to enter the international supply chain for higher value fish. It is investing in landing sites with sanitary facilities, and with cooling, smoking and storage facilities in four key centres – Tombo, Bonthe, Shenge and Goderich. Assuming there is fair access to these facilities for all fishers, and there are no 'informal' charges, there could be a major improvement in the local economies. Other forms of aid, however, like providing loans for boats and equipment fishers lack the income to service (eg. in New Ireland, PNG and also in Fiji), can prove wasteful and unsuccessful. Capacity-building for people, where they stay in fishery-related fields, has had better results. The National Fisheries College in New Ireland, supported by the PNG government, and the Caribbean Fisheries Training and Development Institute (CFTDI) in Trinidad, which benefited from Japanese aid, have both provided such courses for people from across their regions. And in regions like the Caribbean and Pacific with many relatively small states, regional programmes seem to be a good way forward, particularly if there are more scholarships to centres outside the study region, for different types of specialist. In places as diverse as St Lucia, Samoa and Mozambique, fishery management benefits enormously from qualified people who stay for many years at their posts. Capacity building, founded on local institutions which have established their own expertise, is particularly valuable. There may also be scope for more transfer of skills and technology between Commonwealth countries. For example the Seafood Industry Development Company, a parastatal in Trinidad, as well as the CFTDI, could usefully pass on their skills to people outside the Caribbean. Promoting value-added fish projects can lead to more employment, and better earnings for fishing communities.

## Viability of Coastal Communities, Including Non-fish and Tourism Incomes

Community viability rarely depends entirely on fishing and in many of the countries studied few fishers worked full-time or every day. Many also have incomes from small-scale horticulture and agriculture, from tourism and other sources. Income also is a family phenomenon, involving men, women and children. However fish caught for the pot is a direct contribution to a family's food security. Many communities are trying, therefore, to maximise earnings from a variety of sources. This applies to community-based management structures, like the committees running MPAs, as well as to families. They are looking for a balance of revenues, especially where nearshore fisheries are seriously depleted and MPAs must initially implement no-take zones to allow stocks to recover. The viability of coastal communities is also affected by non-economic pressures: demographic growth, housing problems, or the migration of young people looking for more attractive prospects. In some Caribbean states the immigration of foreigners, seeking a sun-loving lifestyle, poses additional challenges to viability. Generally speaking, fishing and fish-based employment cannot guarantee community viability, even where stocks are sustained. Coastal communities need strategies that go beyond fisheries.

## Sustainable Fishing Strategies, Including Marine Protected Areas

From the work reported, here, several recovery and sustainability strategies are evident. Some relate to the technology of fishing – mesh size, for example – and are designed to avoid the capture of juveniles, to recognise the cycle of spawning, maturity and, in some cases, migration. Rules on larger mesh sizes, which allow small fish to escape, have been introduced for instance in Tanzania and, more gradually, in Mozambique. It has been much harder for governments to police the drastic industrial trawling techniques that damage the sea bottom or lead to huge waste in by-catch. By-catch in the valuable prawn fishery in the Mozambique channel can amount to 70 per cent.

There are also strategies of no-take and avoidance. In Tanzania, where one stock was perilously reduced, a fishing company simply stopped catching it. More typically, MPAs work on the basis of no-take zones surrounded by areas where it is permissible to fish with some targeted restrictions. The idea, which has proved successful, is that, if habitat is protected, species such as trochus and conch can rebuild in the safe oasis and then spread out into neighbouring waters. While no-take was initially controversial with fishers in many coastal areas, the success of this strategy now has broad acceptance, provided MPA surveillance stops poachers from penetrating the marine riches in no-take zones.

## Culture and Gender Issues

The influence of culture on small scale fishing is nowhere better illustrated than in the Pacific, where a God-fearing Christianity is linked to chiefly authority and traditions of '*tabu*'. '*Tabu*' means that on sacred occasions, including the funeral and mourning of a chief, fishermen do not go out in their boats for as long as 100 days at a time. This cultural prohibition on fishing may represent an instance of customary practices being tied to behaviours that also make sense for ecological reasons (McIntosh, 2004). And it might even be possible to modernise this approach and to use it for sustainability purposes in the design of MPAs, where the combination of strong chiefly leadership and customary and Christian tradition – even though these are waning in parts of the Pacific – leads to easier acceptance of a long-term approach to fishery sustainability. The study tour came across one case, however, where pressure from an external NGO to adopt a women's equality approach to fisheries management came up against chiefly conservatism on human rights and undermined efforts for marine sustainability.

The role of women in artisanal fishing is, nevertheless, significant. Women often market and process the catch and in some cases own the boats. In Sierra Leone women are responsible for smoking and trading fish and are active and vocal in the union. As recent declines in fish supplies have made it harder to supply inland markets, there have been declines in income and increased stress on community and family cohesion. Trinidad has a Women in Fishing Association in Mayaro, which aims to add value by processing the fish. But this degree of organisation of the roles of women in small fisheries is unusual.

## Potential and Actual Commonwealth Roles

The case studies and study tours reaffirm the value of richly contextual understandings of the policies and practices relating to small-scale fishers and their communities: ecological, social, cultural and historical. They also underscore the crucial influence of capacity –

even small states with highly competent individuals in ministries and fisheries departments cannot do a better job unless they also have the benefits of being integrated into strong regional networks of support and expertise and of being allied to organisations that can share with them the costs of research and training. Government dependency on NGOs and project-based funding to supply these benefits risks discontinuity and is no substitute in the long term for developing capacity in government itself. In a context of limited government capacity and some corruption the engagement of fisherfolk themselves in community-based management and self-regulation pays dividends. Involving fishers in monitoring ecosystems and fish stocks has proven to be an especially promising development. This involvement can also be coupled with provision for complementary livelihood opportunities for fishers. Such provision recognizes the limited educational and financial resources of artisanal fishers. It also and more importantly recognizes the importance of fishing to identity; that it is a way of life not just an occupation. Livelihood diversification should not distract attention, however, from the need to secure sustainable fisheries management (Gillett and van Santem, 2008), and may in fact have perverse outcomes. Recognition of the need for better baseline data on small-scale fisheries has not been matched by a proper accounting of the contributions they make to national and local economies. And there has been almost no recognition of the social and security costs of the collapse of fisheries in countries that lack social policy to mitigate the effects.

The Commonwealth Fisheries Programme has elicited appreciation for the Commonwealth contributions already made in support of small fisheries, whether by individual Commonwealth nations, especially more developed states, or through the collaborative work funded by the Commonwealth Secretariat to advance fisheries planning in the Pacific and Caribbean (see p. 228). The project also makes it clear, however, that the very considerable wealth of Commonwealth expertise could be harnessed even more effectively in future to address problems of over-fishing and depletion.

Specific suggestions are for:

- A Commonwealth Ministerial task force to strengthen the monitoring of EEZs and reduce depletion;
- A combined Commonwealth and FAO conference on fishery matters;
- An annual Commonwealth assessment of marine fish stocks with particular reference to small-scale fisheries;
- A lobbying effort by Commonwealth agencies for regional funding streams to build the capacity of local NGOs, and support community-specific MPAs, restocking and fund-raising initiatives;
- The strengthening of academic and civil society networks to exchange information and best practice;
- An analysis by member states of the economic, social and community value of subsistence artisanal fishing;

- Special support from the Commonwealth Fund for Technical Cooperation for fishers' organisations in areas like refrigeration, added-value processing, organisation, marketing and transport;
- The systematic extension and networking of MPAs;
- Continuing support for regional organisations, such as SADC, CARICOM and the Secretariat of the Pacific Community, as well as the Regional Fisheries Management Organisations, in the area of sustainable fisheries management; and
- A request to Commonwealth donors to prioritise small states in awarding monies for training, scholarships, EEZ surveillance and boundary delimitation.

## References

Andrews, N. and Evans, L. (2009) *Approaches and Frameworks for Management and Research in Small-scale Fisheries in the Developing World*. The WorldFish Center Working Paper 1914 (Penang: The WorldFish Center).

DFID, World Bank and FAO (2007) *Sierra Leone: A proposed Fisheries Sector Strategy Paper* (FSSP). Prepared by Neiland, A.E., Arbuckle, M., Bodiguel, C., Bostock, T., Campbell, J., Gronnevet, L., Kaindaneh, P., Seisay, M., and Williams, L.O. (London: UK Department for International Development, Washington: World Bank, and Rome: United Nations Food and Agricultural Organisation).

EJF (2009) *Dirty Fish – How EU Hygiene Standards facilitate illegal fishing in West Africa* (London: Environmental Justice Foundation).

Fairweather-Morrison, Imani (2006) 'Consideration of socio-economic and demographic concerns in fisheries and coastal area management and planning in Belize. Chapter 1. Case Study: Belize'. In: Tietze, U.; Haughton, M.; Siar, S.V. (Eds) *Socio-Economic Indicators in Integrated Coastal Zone and Community-based Fisheries Management – Case Studies from the Caribbean*. FAO Fisheries Technical Paper No. 491 (Rome: FAO) pp. 5–27.

FAO (2007). Making global governance work for small-scale fisheries. *New Directions in Fisheries – A Series of Policy Briefs on Development Issues*, No. 09, Rome, 16 pp., also available from *http://www.sflp.org.briefs/eng/policybriefs.html*

FAO and World Fish Center (2009) *Small Scale Capture Fisheries – A Global Overview with Emphasis on Developing Countries: a preliminary report of the Big Numbers Project* (Rome: FAO and Penang: World Fish Center).

Gillett, R. and G. van Santen (2008) *Optimizing Fisheries Benefits in the Pacific Islands: Major Issues and Constraints* (Washington, D.C.: World Bank).

Leisher, Craig Pieter van Beukering and Lea M. Scherl (2007) *Nature's Investment Bank. How Marine Protected Areas Contribute to Poverty Reduction*, 43pp. The Nature Conservancy, accessed at: *http://www.nature.org/initiatives/protectedareas/files/mpa_report.pdf* 46pp.

McIntosh, Alistair (2004) *Soil and Soul. People versus Corporate Power* (London: Aurum Press).

Pacific Islands Forum Secretariat (2007) 38th Pacific Islands Forum, Nuku'Alofa, Tonga 16–17 October 2007. Forum Communiqué, Annex A. Vava'u Decisions on the Pacific Plan p. 8.

Tawaki, Alifereti, Parks, John, Radikedike, Pio, Aalbersberg, Bill, Vuki, Vaikila and Salafsky, Nick (2001) Harvesting Clams and Data. Involving local communities in monitoring: A case in Fiji. *Conservation Biology in Practice* 2 (4) pp. 32–35.

Thorpe, Andy, David Whitmarsh, Ernest Ndomahina, Andrew Baio, Miatta Kemokai and Thomas Lebbie (2008) 'Fisheries and failing states: The case of Sierra Leone'. *Marine Policy*, doi:10.1016/j.marpol.2008.09.002.

Veitayaki, Joeli (1995) *Fisheries Development in Fiji: The Quest for Sustainability*. Suva, Fiji. Insitute of Pacific Studies, The University of the South Pacific.

Watts, Nicholas and Geoffrey Wandesforde-Smith (2006) 'The Law and Policy of Biodiversity Conservation in the Caribbean: Cutting the Gordian Knot'. Guest Editors' Introduction to Watts, Nicholas and Wandesforde-Smith, Geoffrey (eds), *Journal of International Wildlife Law and Policy*, 9:3 Special Issue *The Law and Policy of Biodiversity in the Caribbean*.

## Notes

1 Details of the two year programme, including full reports, are publicly available on the website: *www.commonwealthfisheries.org*

2 See FAO (2007); Andrews and Evans (2009); FAO and World Fish Center (2009).

3 For consideration of the regional approach to biodiversity in the Caribbean, see Watts and Wandesforde-Smith, 2006).

4 There is, nonetheless, conflict within the fishing industry in Fiji along ethnic lines and it has proved divisive in the country over the last two decades. The main friction is between the Indo-Fijian owners of larger, longer-range boats that fish out of Suva and the bigger harbours of Fiji and the smaller boats fished by indigenous Fijians. The commercial Indo-Fijian fishers are better able to purchase rights to fish inside traditional *qoliqolis* from the chiefs.

5 Dealing with the artisanal fishery issues noted here is not, however, a straightforward process of drawing lessons from particular instances and transferring them to other contexts, especially when interactions between customary and modern institutions are concerned. In Samoa, for example, the chiefly tradition includes the right of the chief to banish poachers from the community. In the Fiji case study, however, when a chief confiscated fish from a poacher in his *qoliqoli*, he was arrested on larceny charges.

# 3 WTO Negotiations on Fisheries Subsidies: A Critical Issue for Commonwealth Countries

Anja Von Moltke

## Introduction

Fish stocks around the world are facing an unprecedented crisis of depletion. A recent report issued by the World Bank and FAO has shown that the annual global loss in wealth caused by the depletion of fish stocks is 'conservatively estimated in the order of $50 billion' (World Bank and FAO, 2008, p. 38). Today, there is broad agreement that inappropriate subsidies are a real part of the problem. Estimated to be worth US$15–35 billion annually, fisheries subsidies come in a wide variety of forms – including direct cash grants, tax breaks, loan guarantees, and even the provision of goods and services (Milazzo, 1998; WWF, 2001; UNEP, 2004a; Sumaila and Pauly, 2006). They are applied for an equally broad variety of purposes, ranging from the direct promotion of expanded fishing capacity and productive effort to support for early retirement and the reduction of fishing fleets.

Although properly designed fisheries subsidies can help achieve responsible fishing practices, economists and fisheries experts agree that many fisheries subsidies contribute to overfishing. The majority of fisheries subsidies are granted by a handful of economically powerful nations.[1] Although details of subsidies often remain hidden, there is little doubt that subsidised fleets maintain an unfair advantage in the race for dwindling fish stocks. Subsidies thus distort competition, mainly to the disadvantage of those that do not subsidise, while depleting the stocks on which all fishers depend.

As a consequence, fisheries subsidies have become a topic of international negotiations, most notably at the World Trade Organization (WTO). This paper begins by reviewing the importance of the fisheries sector for Commonwealth countries and their stance on fisheries subsidies. It then explains the origins and history of the WTO negotiations on fisheries subsidies and analyses the key issues within the negotiations that still have to be solved in order to reach an international agreement on fisheries subsidies. The paper concludes that engagement in the negotiations provides Commonwealth countries with a unique opportunity to improve the competitiveness of their own industries in international trade while contributing to the preservation of their fisheries resources.

## Fisheries Subsidies in Commonwealth Countries

Many of the Commonwealth member states are fishing nations, even though of highly diverse nature. Fisheries operations range from subsistence to industrial, and supply between approximately 1 per cent and 55 per cent of total protein intake in Commonwealth countries. In aggregate, the 53 Commonwealth nations produce approximately 13 per cent of the global marine fish catch (including crustaceans) (derived from WRI, 2006). For many Commonwealth members – as for developing countries generally – fish constitutes a major source of foreign exchange earnings. The value of net exports of fish from developing countries by far exceeds those of coffee, rubber, and cocoa (FAO, 2008). For instance, the value of fish exports amounts to 26 per cent for the Seychelles, 14 per cent for Vanuatu, 8.5 per cent for the Maldives, and 6 per cent for Solomon Islands.[2] In India, the export value of fish, crustaceans, molluscs and preparations thereof rose from USD$1306.7 million to USD$1756.2 million between 2003 and 2006 (UN Comtrade, 2008).

Approximately 38 per cent of all Commonwealth countries are Small Island Developing States (SIDS). These produce only 8 per cent of the Commonwealth marine fish catch (derived from WRI, 2006), but are largely dependent on fish for their own food security, livelihoods and foreign exchange earnings. Just in the Eastern Caribbean states, for instance, fisheries contribute over EC €138 million to the region's GDP (OECS, 2009). In South Africa, fishing only constitutes 1 per cent of total GDP, but in some coastal regions, such as the Western Cape, it is the main source of income (FAO, n.d.).

Fisheries are also important to many Commonwealth members through the practice of selling foreign access to their EEZ waters. Access agreements which can be associated with subsidies granted by flag states to their distant water fleets (see below) – bring substantial foreign exchange earnings to a number of small island and coastal Commonwealth countries (Campling and Havice, 2007; UNEP, 2008). At the same time, they have led to controversy over their ultimate benefits to coastal host countries, where they frequently appear associated with stock depletion. Since they are often negotiated in secret, and without adequate knowledge of fishery conditions, access agreements potentially result in the transfer of overcapacity from developed countries to developing countries already facing overexploitation (Mwikya, 2006).

Commonwealth countries have taken a variety of positions at the WTO fisheries subsidies negotiations, reflecting the differences among them in their dependence on the fisheries sector, the state of their fish stocks, their trade and competitiveness concerns, and, of course, their current practices related to subsidies. Developed Commonwealth members (e.g. Australia, Canada, New Zealand, UK) also have interests quite different from those of developing members (e.g. Pacific Island countries, African countries). But even within these groups, positions differ substantially. Australia and New Zealand, for example, are part of a group called 'Friends of Fish'[3] advocating a strong prohibition on fisheries subsidies with only few exceptions. Canada, on the other hand, tends to join the EU and Japan in calling for prohibition of a more limited list of subsidies.

Some of the subsidising developing countries, such as India, call for the right to continue using subsidies without any conditionality imposed. Others, such as some for the Pacific Island countries hardly subsidise their fisheries, but would yet like to maintain the right to subsidise in the future in order to further develop their industries – within sustainable limits.

### Box 1

**New Zealand's stance towards subsidies**

*'Around 90 percent of our catch is exported, making New Zealand heavily dependent on world markets. [...] Catches in some Northern Hemisphere fisheries have fallen recently, bringing hard times for fishers. Some countries have helped their industries through subsidies. However, these make it cheaper for the companies to catch fish, which is unfair for New Zealand exporters. Some subsidies are for building more fishing boats, which can lead to overfishing. So the government is working in the World Trade Organization to end fishing subsidies.'*

Ministry of Fisheries, New Zealand (2007)

**Submission to the WTO by SVEs (Barbados, Cuba, Dominica, Dominican Republic, El Salvador, Fiji, Honduras, Jamaica, Mauritius, Papua New Guinea, St Lucia, St Vincent and the Grenadines and Tonga)**

*'The Small, Vulnerable Economies (SVEs) have actively participated in the negotiations on disciplines on fisheries subsidies with the aim of contributing to fulfilling the Ministerial mandates on fisheries subsidies negotiations and to ensure that there is an appropriate balance found between sustainability concerns and the development priorities of developing countries, especially for those countries which have little or no impact on over fishing and over capacity.'*

TN/RL/W/226/Rev.4

As can be seen from this short overview, Commonwealth countries face different situations related to the nature and condition of their fisheries, their trade interests and their positions towards subsidies. At the same time, there are overarching issues that are of interest to many of them: securing livelihoods, increasing foreign exchange earnings and providing food security. There is a growing realisation amongst all parties now that these aims can only be achieved in the long run if fish stocks are kept at sustainable levels.

## WTO Negotiations – A Brief History

Fisheries subsidies first came onto the international agenda in the 1990s, following studies conducted by the FAO, UNEP, WWF and the World Bank, among others. The studies revealed a significant level of subsidisation, and suggested strong links to fisheries depletion (Milazzo, 1998; WWF, 2002; Schrank, 2003; UNEP, 2003a).

At the WTO, fisheries subsidies were originally addressed in the Committee on Trade and Environment (CTE), starting in 1997. At the WTO Seattle Ministerial Conference in 1999, a group of countries, the so-called 'Friends of Fish' put forward proposals aimed at launching negotiations on a new agreement that would specifically discipline fisheries subsidies that distort trade, harm the environment, and impede development.

As a result, the Ministerial Conference in Doha in 2001 launched negotiations aiming to 'clarify and improve WTO disciplines on fisheries subsidies, taking into account the importance of this sector to developing countries' (Para. 28 of the Doha Declaration). The mandate also made explicit reference to the aim of 'mutual supportiveness of trade and environment' (cross reference in Para. 28 to Para. 31), which clearly linked the negotiations on subsidies and environmental issues.

Since 2001, the negotiations on fisheries subsidies have taken place in the WTO Negotiating Group on Rules (NGR), which is under the authority of the WTO Trade Negotiations Committee (TNC) and also addresses antidumping and regional trade agreements. While the CTE remains formally active on fisheries subsidies, the central work of the WTO on the subject has shifted to the NGR.

When world leaders met at the 2002 Johannesburg World Summit on Sustainable Development a few months after the Doha Ministerial Conference, they listed successful conclusion of the WTO fisheries subsidies talks as a top priority for achieving sustainable fisheries (see Box 2).

### Box 2: Call from the World Summit on Sustainable Development

'To achieve sustainable fisheries, the following actions are required at all levels:

[...] Eliminate subsidies that contribute to illegal, unreported and unregulated fishing and to over-capacity, while completing the efforts undertaken at the World Trade Organization to clarify and improve its disciplines on fisheries subsidies, taking into account the importance of this sector to developing countries [...].'

Report of the World Summit on Sustainable Development, Johannesburg, 26 August–4 September 2002, 31(f)

By the spring of 2004, negotiations had reached a stage where it was no longer a question of *whether* but of *how* international cooperation to reform fisheries subsidies should move forward.[4] Discussion then focused on whether to adopt a 'bottom up' approach to identifying and defining prohibited subsidies – as proposed by Japan, Korea and the EU – or to start with a general prohibition and focus the negotiations on exceptions thereto. This latter 'top-down' approach was promoted by the 'Friends of Fish'.

These developments – along with signs that leading developing countries would support robust fisheries subsidies that included solid 'special and differential treatment' language – paved the way for a breakthrough agreement at the December 2005 'mid-round' WTO ministerial in Hong Kong. While diplomats struggled in vain to adopt 'modalities' on other key Doha Round issues, ministers issued a revised negotiating mandate that called for an enforceable ban on fisheries subsidies that contribute to overcapacity and over-fishing, along with a strong call for special and differential treatment for developing countries (see Box 3).

### Box 3: The Hong Kong Mandate

'We [ministers]. . . recall our commitment at Doha to enhancing the mutual supportiveness of trade and environment, note that there is broad agreement that the Group should strengthen disciplines on subsidies in the fisheries sector, including through *the prohibition of certain forms of fisheries subsidies that contribute to overcapacity and over-fishing*, and call on Participants promptly to undertake further detailed work to, inter alia, establish the nature and extent of those disciplines, including *transparency and enforceability*. Appropriate and effective *special and differential treatment* for developing and least-developed Members should be an integral part of the fisheries subsidies negotiations, taking into account the importance of this sector to development priorities, poverty reduction, and livelihood and food security concerns ...'

Hong Kong Declaration, Annex D, § I.9 (emphasis added)

The Hong Kong commitment made headlines around the world, and took the talks to a new level of intensity.[5] Over the next two years, technical proposals on a range of fisheries subsidies topics were submitted by various WTO delegations and debated intensely.[6] The fisheries subsidies talks most likely benefited from a delay experienced in the general negotiations, which allowed continuing formal and informal dialogue on key issues.[7] Generally, it can be noted that negotiations on these issues benefited from the close involvement and initiative taken by UNEP and several NGOs, such as WWF, ICTSD, and Oceana.

As a result, at the end of November 2007 the chairman of the NGR released a first 'chair's draft' of proposed WTO rules governing, among other issues, fisheries subsidies (TN/RL/W/213). The draft legal text included the following elements:

- A prohibition of a broad range of directly capacity – or effort-enhancing fisheries subsidies, as well as any subsidies affecting fishing on 'unequivocally overfished stocks';
- Exemptions of several specific classes of subsidies from the prohibition (e.g. for vessel safety or reducing fishing capacity);
- A condition that basic fisheries management systems be in place for those subsidies that remain permitted;
- The possibility for developing countries to use most prohibited subsidies, subject to fisheries management;

- A mechanism for involving the FAO in the review of measures taken to fulfil fisheries management criteria; and
- Strengthened WTO notification rules for fisheries subsidies.

Although the document received significant support as a basis for negotiations, it was clear that many elements raised controversy and uncertainty. In December 2008, the Chair recognised the need to revisit some fundamental questions and issued a new document containing a 'roadmap' to guide further talks (TN/RL/W/236). This roadmap is based on his initial 'draft text' and highlights the need particularly to focus on the following issues, discussed in more detail in the next section:

- What should be the scope of an eventual prohibition on certain classes of fisheries subsidies?
- What 'sustainability criteria' should be placed as conditions or limits on fisheries subsidies that remain permitted?
- What should be the scope and nature of 'special and differential treatment' (S&DT) for developing countries?
- What rules and mechanisms should be put in place to ensure transparency and enforcement?

## Key Issues for the Negotiations

### The scope of the prohibition

The Hong Kong mandate calls for a WTO prohibition on 'certain forms of fisheries subsidies that contribute to overcapacity and over-fishing'.[8] Most experts – and many governments – agree that direct subsidies to the capital costs of fishing enterprises contribute to overcapacity and overfishing, and thus that subsidies such as those to vessel construction, outfitting, or modification, should be prohibited. The issue of operating costs, such as fuel subsidies, is more controversial. While some argue they should not be included within the ban, others feel that they clearly contribute to increased fishing pressure and thus to overfishing and should be prohibited.

Another set of issues surrounds the treatment of subsidies that make a less direct contribution to overcapacity and overfishing. For example, subsidies to port infrastructure, to initial processing, or to marketing or price supports are often considered by economists as having potential impacts on the economic incentives affecting fishing enterprises. Still, it is widely recognised that the 'links to depletion' may be less compelling in such cases.

Another question affecting the ultimate scope of a prohibition is the set of exceptions that may accompany it. A general consensus exists that some kinds of subsidies should be excluded from an eventual ban, such as subsidies for improving vessel safety, for relief from natural disasters, and for vessel decommissioning.

An early submission by a number of SIDS (Antigua & Barbuda, Belize, Fiji, Guyana,

Maldives, Papua New Guinea, Solomon Islands, St Kitts & Nevis), however, had proposed to not only exempt, but exclude from any new disciplines: (i) Access Fees and Development Assistance; (ii) Fiscal Incentives to Domestication and Fisheries Development; and (iii) Artisanal Fisheries (TN/RL/W/136). Their concerns related to these issues have later been taken up in the discussion on new disciplines, particularly under Special and Differential Treatment (S&DT) related items.

The negotiations on the scope of the ban are considered to be the most important elements of the negotiations, since the outcome will, to a large part, determine the impact and effectiveness of the new disciplines.

## Sustainability criteria

The conditions prevailing in a fishery can have a significant impact on the potential for subsidies to do harm, as shown in a UNEP analysis of the environmental impact of different types of subsidies under different management conditions (see Figure 1) (UNEP, 2004a; 2004b; 2004c).

The study showed that under most real-world conditions, i.e. less than perfect management schemes and fully exploited fish stocks, subsidies can have very harmful environmental effects. While it is hoped that new WTO rules will sharply curtail the use of the most harmful kinds of fisheries subsidies, it is clear that various forms of public support for the fishing industry will continue in the future. It is also clear that where fishing fleets are well below full capacity, and where fisheries management systems are strong and effective, the potential for fisheries subsidies to cause harm is significantly lower.

Recognising this fact, many WTO members have indicated support for rules that employ sustainability criteria based on these factors as preconditions on the use of permitted fisheries subsidies.

In light of the unfortunate fact that many of the world's fisheries are not yet sustainably managed, the need for sustainability criteria in WTO fisheries subsidies rules has raised difficult questions about the level of rigour that such criteria can reasonably impose.[9]

The discussion has drawn on international norms and standards emanating from the UN Code of Conduct for Responsible Fishing. The Code of Conduct requires fisheries to be managed to achieve long term maximum sustainable yields through the application of three fundamental elements of management:

- Assessment of fish stocks and fleet capacity;
- Limitation of fishing and fishing capacity through regulatory controls; and
- Surveillance and enforcement of regulatory limits.

A recent UNEP-WWF publication entitled *Sustainability Criteria for Fisheries Subsidies: Options for the WTO and Beyond* demonstrates that the use of such criteria can be tailored for application in the WTO as well as within a national context (UNEP/WWF, 2007).

**Figure 1: Impact of eight categories of fisheries subsidies on fish stocks**

| | Over-capacity | Full capacity | Less than full | Over-capacity | Full capacity | Less than full |
|---|---|---|---|---|---|---|
| Fisheries Infrastructure | H | H | NH | H | H | NH |
| Management Services | NH | NH | NH | NH | NH | NH |
| Access to Foreign Waters | H | H | NH | H | H | NH |
| Decommissioning | PH | PH | — | H | PH | — |
| Capital Costs | H | H | H | H | H | H |
| Variable Costs | H | PH | PH | H | H | PH |
| Income Supports | PH | PH | PH | H | H | PH |
| Price Supports | H | H | PH | H | H | PH |

NH = Not harmful
PH = Possibly or probably harmful
H = Harmful
– = Not applicable

Source: Modified from UNEP (2004a), *Analyzing the Resource Impact of Fisheries Subsidies: A Matrix Approach*. UNEP, Geneva.

It is clear that scientific information relating to fisheries management and health of fish stocks would be important for the practical application of sustainability criteria within new fisheries subsidies disciplines. Many delegations have noted though that the provision of such information may raise questions that go beyond the competence of the WTO and thus may require some structured involvement by an organisation with fisheries expertise, such as the FAO or Regional Fisheries Management Organisations.

## Special and differential treatment

From the outset of the negotiations, the Doha mandate has put special emphasis on the importance of the fisheries sector to developing countries.[10] This mandate was substantially strengthened in Hong Kong, when ministers included strong and specific instructions regarding Special and Differential Treatment (S&DT) in new fisheries subsidies disciplines (see Box 2).

Throughout the S&DT discussion, many delegations have voiced the need to preserve the 'policy space' necessary to encourage development of their domestic fisheries and fish processing sectors. As mentioned above, in many developing countries, and especially in SIDS and other countries with large coastal areas, fishing contributes significantly to food security. Further, the fisheries sector provides employment, both directly in fishing activities and in dependant sectors. In addition, fish and fish products have become an important source of foreign exchange, with 50–60 per cent of fish internationally traded coming from developing countries.

The discussion on S&DT is thus of core importance to those countries wishing to allow their economies to diversify and to preserve fish as a resource base for their economies. An underlying theme of the WTO discussions therefore has been that policy space is necessary for developing countries, but that S&DT should not simply be a 'blank cheque' for the use of subsidies without regard to the impacts on sustainability. For instance, small vulnerable economies (SVE) countries have emphasised that 'the SVEs have not requested a substantial roll-back of the fisheries management principles established in Article V' (TN/RL/W/226/Rev.4). India, Indonesia and China, also acknowledge the need for effective management systems, but argue that developing countries need to be able to establish these at their own pace (TN/RL/GEN/155/Rev.1). Finding an appropriate balance between rules that are too lax and rules that are impossibly burdensome will be critical to the ultimate success of the talks.

Another important question yet to be fully resolved relates to 'artisanal' or 'small scale' fisheries (UNEP, 2005a; UNEP, 2005b). 'Artisanal fishing refers broadly to small, underdeveloped, and often severely impoverished fishing communities whose immediate survival depends on their ability to continue benefiting from local fisheries that in many cases are centuries old' (UNEP, 2005a). The definition in the WTO context is subject to debate. Questions have for instance been raised on whether a limit on the length of a fishing boat could be used as a criterion for determining whether it can be considered 'artisanal' and whether it be exempted from the rules.

An issue of special importance to many SIDS, is the treatment of subsidies that are related to rights of access given to foreign fleets for fishing in developing country waters. Mostly, access agreements are concluded between developing coastal and island states and so-called distant water fishing nations, such as the EU, US and Japan. For SIDS, the payments transferred under such access agreements can contribute up to 50 per cent of GDP (Mwikya, 2006). At the same time, access agreements have come under international scrutiny as a result of their negative impacts both on fisheries resources, including associated ecosystems, and on international markets (UNEP, 2008). Case studies in Senegal have shown, for instance, how distant water fishing nations, facing overexploitation and fisheries collapse in their own waters, have transferred their problem of overcapacity to distant waters via bilateral access agreements (UNEP, 2003b). Mwikya (2006) also argues that only 10 per cent of the employment and value-added generated through these access agreements go to the host country.

The WTO discussions on 'access-related' subsidies have at times been intense, and have resulted in an apparent consensus in favour of strictly limiting the definition of the 'subsidy' element in the access relationship. Specifically, it has been proposed to exclude access fees paid by one government to another from the definition of 'fisheries subsidy', while treating the onward transfer of access rights as a possible subsidy. In combination with strengthened capacity and funding for adhering to sustainability criteria, this compromise could ensure that developing countries continue benefiting from the financial transfers included in access agreements, while at the same time improving the terms on which access is granted and potentially increasing the competitiveness of their own fleets.

Further important questions about the scope and content of S&DT for fisheries subsidies remain unsettled. These include for example questions about the territorial limits of S&DT, such as restricting the application of S&DT to subsidies for fishing within the territorial waters or Exclusive Economic Zones of subsidising countries. Moreover, in light of the very large differences in development and economic power among developing country fishing industries, S&DT for fisheries subsidies also raises the politically volatile question of whether rules for S&DT can differentiate between developing countries with fisheries industries at different stages of development.

### Transparency and enforcement

Finally, the Hong Kong mandate specifically calls for WTO fisheries subsidies rules to address issues of transparency and enforcement (see Box 3). These issues, sometimes treated as a technical afterthought, are fundamental to the ultimate effectiveness of new WTO fisheries subsidies rules, particularly as very poor transparency is endemic to fisheries subsidies programs. Estimates are that approximately 90 per cent of fisheries subsidies are not notified (Schorr 2004).

It is evident that current WTO rules governing subsidies notification have no 'teeth' – i.e., there are no legal consequences when governments fail to notify. A prohibition of those subsidies that are not notified would be an important element of new rules. In addition, given the importance of sustainability criteria for permitted subsidies, it would be helpful to include information about stock and management conditions of subsidised fisheries to allow for an evaluation of whether sustainability criteria have been met.

The need for improved transparency in fisheries subsidies is universally acknowledged. However, governments still need to be further encouraged to impose on themselves WTO rules that are significantly strengthened in terms of transparency and enforcement.

## The Challenges Ahead

In the face of multiple crises, including the global economic, financial and environmental crisis, urgent attention should be given to investing financial resources more wisely and more sustainably and to providing greater transparency on financial transfers including subsidies. Although governments themselves are responsible for ensuring that their fisheries policies do not conflict with sustainable development, the WTO undoubtedly has a major role to play.

Despite all the problems that the overall Doha Round is facing, significant substantive and political progress has been made over the past years on the reform of fisheries subsidies. However, there are many questions that remain for WTO negotiators to resolve in order to deliver a successful outcome. These questions are neither trivial nor insurmountable – but require urgent solutions. Significant efforts will have to be made to find agreement on the scope of the ban, the management conditions that will accompany those subsidies that will remain permitted, special and differential treatment provisions as well as modalities for enhancing transparency.

A triple win outcome of the negotiations for trade, environment and development is possible, but only if a healthy synergy between environmental protection and development is found. It will therefore be essential to consider mechanisms that can assist developing countries in enhancing their capacity to put in place effective fisheries management systems.

Apart from benefiting from such technical assistance that should go with the reform of fisheries subsidies, it is clear that many developing countries would substantially benefit from lower levels of subsidies globally, making their own fleets and fishing industries more competitive.

The threat facing the world's marine resources is not only an environmental challenge, it has crucial bearing on food security, employment and livelihoods. Participating actively in the fisheries subsidies negotiations will give an opportunity to all stakeholders from Commonwealth countries to bring local realities into the debate and to support an outcome that will directly promote sustainable development.

## References

Campling, L. and Havice, E. (2007). *The Pacific Islands, The Global Tuna Industry and the International Trade Regime – A Guidebook* (Honiara: Forum Fisheries Agency).

FAO (2006) Fish exports by developing countries help combat hunger, but better management needed: FAO proposes guidelines for responsible fish trade at meeting in Spain. *FAO News*, 30 May 2006. *http://www.fao.org/newsroom/en/news/2006/1000301/index.html*

FAO (2008) *State of the World Fisheries and Aquaculture* (Rome: FAO).

FAO (n.d.) *Fishery and Aquaculture Country Profile – South Africa. http://www.fao.org/fishery/countrysector/FI-CP_ZA/en*

Milazzo, M. (1998) *Subsidies in world fisheries: A re-examination*, World Bank Technical Paper No. 406.

Ministry of Fisheries, New Zealand (2007) Market Access. *http://www.fish.govt.nz/en-nz/Publications/State+of+our+fisheries/International+Fisheries/Market+access.htm?&MSHiC=65001&L=10&W=negotiations+world+trade+organization per cent20&Pre= per cent3Cspan per cent20class per cent3d per cent27SearchHighlight per cent27 per cent3E&Post= per cent3C/spanper cent3E*

Mwikya, S.M. (2006) *Fisheries Access Agreements: Trade and Development Issues*. (Geneva: ICTSD)
OECS (2009) *http://www.oecs.org/Press/News_OECS_Secretariat_helps_officials_to_craft_positions_on_Fisheries_Subsidies.html*
Report of the World Summit on Sustainable Development *http://www.unmillenniumproject.org/documents/131302_wssd_report_reissued.pdf*
Schorr, D.K. (2004) *Healthy Fisheries, Sustainable Trade – Crafting New Rules on Fishing Subsidies in the World Trade Organization* (WWF).
Schrank, W.E. (2003) *Introducing Fisheries Subsidies* (Rome: FAO).
Sumaila, R. and Pauly, D. (2006) *Catching More Bait: A Bottom-Up Re-Estimation of Global Fisheries Subsidies* (University of British Columbia Fisheries Centre).
UN Comtrade (2008) *http://comtrade.un.org/pb/CountryPages.aspx?y=2006*
UNEP (2003a), Gareth Porter *Fisheries Subsidies and Overfishing – Towards a Structured Discussion* (Nairobi).
UNEP (2003b) *Fisheries Subsidies and Marine Resource Management: Lessons learned from Studies in Argentina and Senegal.*
UNEP (2004a) *Analyzing the Resource Impact of Fisheries Subsidies: A Matrix Approach.*
UNEP (2004b) Chair's Summary, UNEP Workshop on Fisheries Subsidies and Sustainable Fisheries Management, 26–27 April 2004 (*http://www.unep.ch/etb/events/events2004/FisheriesMeeting/FinalChairsSummary.pdf*).
UNEP (2004c) *Incorporating Resource Impact into Fisheries Subsidies Disciplines: Issues and Options.*
UNEP (2005a) *Artisanal Fishing: Promoting Poverty Reduction and Community Development Through New WTO Rules on Fisheries Subsidies*. An Issue and Options Paper (Nairobi).
UNEP(2005b) *Issues and Options Papers on Special and Differential Treatment.*
UNEP (2008) Marcos Orellana, *Towards Sustainable Access Agreement, Issues and Options at the World Trade Organization.*
UNEP/WWF (2007) *Sustainability Criteria for Fisheries Subsidies: Options for the WTO and Beyond.*
World Bank and FAO (2008) *The Sunken Billions: The Economic Justification for Fisheries Reform* (Washington DC, USA: World Bank Publishing).
WRI (2006). *Coastal and Marine Ecosystems – Variables. http://earthtrends.wri.org/text/coastal marine/variables.html*
WTO submissions (TN/RL/GEN/155/Rev.1, TN/RL/GEN/156, TN/RL/W/136, TN/RL/W/213 Annex VIII, TN/RL/W/226/Rev.4, TN/RL/W/235, TN/RL/W/236).
WWF (2001). *Hard Facts, Hidden Problems: A Review of Current Data on Fishing Subsidies.*
WWF (2002). *Turning the Tide on Fishing Subsidies – Can the World Trade Organization Play a Positive Role?*

## Notes

1 Two leading efforts have offered slightly different lists of the leading subsidisers: a 2001 study by WWF and a 2006 report by the Fisheries Centre at the University of British Columbia (see references). The latter suggests that developing countries provide half of all fisheries subsidies, but still much less per country than the major economies. Both studies have been subject to some controversy due to the lack of data availability and comparability.

2 Average value for 2002–06 as a percentage of 2007 GDP, based on World Bank (2008).

3 At various times, active members of the Friends of Fish coalition have included Argentina, Australia, Chile, Ecuador, Iceland, New Zealand, Norway, the Philippines, Peru, and the USA.

4 See UNEP workshop report, 2004b: *http://www.unep.ch/etb/events/FishMeeting2004.php*

5 See *http://www.unep.ch/etb/events/UNEPWWFside_eventHK.php*

6 For a good record of these negotiations, see e.g. ICTSD Bridges 2006–2009. For a compilation of the WTO submission see *http://www.unep.ch/etb/areas/WTOSubmissionsFish.php*

7 See, for example, the output of technical workshops co-sponsored by UNEP (*http://www.unep.ch/etb/events/Fish_events.php*).

8 There is a broad understanding now that the ban will focus only on subsidies affecting 'marine wild capture fishing'.

9 The chair's summaries of several UNEP-WWF workshops (2006, 2007, 2008) provide a good overview of this discussion (available on the UNEP website as indicated above).

10 See for example *http://www.gtz.de/en/themen/laendliche-entwicklung/25020.htm and http://eur-lex.europa.eu/LexUriServ/LexUriServ.do?uri=COM:2000:C724:FIN:EN:PDFb*

# 4 The Law of the Sea: A Commonwealth Perspective

Richard Barnes

## Introduction

The law of the sea provides the primary legal context for the regulation of the world's ocean spaces. Both the global nature of subject matter and the global interest in ensuring the proper management of oceanic spaces and resources require an international legal framework of rules. Of course, the law of the sea, as a specialist field of international law, deals in the first instance with the rights and duties of states and, as in the case of international law more generally, its effectiveness is dependent upon the extent to which states implement their international obligations within domestic legal orders. Any analysis of how the law of the sea impacts upon the regulation of fisheries must consider the provisions of the United Nations Convention on the Law of the Sea 1982.[1] This chapter provides a synopsis of the key features of the Convention, noting how its general provisions must shape any initiatives to address current failures in fisheries management in the Commonwealth. It then briefly evaluates its specific provisions concerning fisheries regulation, indicating some of the key weaknesses of the current regime. Since the adoption of the Convention a number of measures have been implemented that aim to address the Convention's shortcomings. These are noted, before some brief concluding observations on the current state of play in international fisheries regulation are offered.

## The Law of the Sea Convention

The United Nations Convention on the Law of the Sea 1982 provides a framework for the regulation of the world's oceans. Indeed, its 320 Articles and 9 Annexes have been hailed as a constitution for the oceans, indicating its fundamental regulatory role. Commonwealth states have played key roles in the regime and the Convention's genesis can be traced to the speech of Maltese Ambassador Arvid Pardo to the General Assembly in 1967. His advocacy of the common heritage principle deeply influenced the resultant legal framework. Kenya was influential in advancing the Exclusive Economic Zone (EEZ) concept through the Yaoundé Declaration 1972 (Houston Lay *et al.*, 1973). The Convention was then opened for signature in Montego Bay, Jamaica. The product of lengthy negotiations at the third United Nations Conference on the Law of the Sea between 1973 and 1982, the Convention eventually entered into force on 16 November 1994. As of March 2009 it had 157 parties, including the European Community. 37 states, including the United States, are not party to the Convention. Not all Commonwealth states have ratified or acceded to the Convention (see Appendix I). Much of the Convention is binding upon all states by virtue of customary international law, regardless of whether or not they are states parties, but wider participation in the Convention and its implementing agreements would provide greater certainty and consistency in the law.

The Convention sets the limits of legal authority to coastal states through a series of maritime zones. These include the territorial sea, which is a 12 nautical mile (nm) band of waters contiguous to the coastline, within which coastal states enjoy full sovereignty subject to third states rights of innocent passage (Article 2–32). Next, a contiguous zone between 12 and 24 nm may be claimed, within which coastal states may enforce certain customs, immigration and sanitary laws (Article 33). In the context of fisheries, the most important zone is the EEZ, which may be drawn up to 200nm from the coastal states baselines (Article 57). Within this zone, coastal states enjoy exclusive rights to exploit and manage living and non-living natural resources (Article 56). In addition, a coastal state enjoys certain jurisdictional rights over the continental shelf up to a limit of 200 miles or, in some circumstances, beyond this limit to the outer edge of the continental margin as defined in the Convention (Article 76). Here coastal states enjoy sovereign rights over the mineral resources, and, somewhat anomalously, sedentary species, such as mussels or crabs (Article 77(4)). Beyond the limits of national jurisdiction, the remaining areas of ocean space are the high seas, which pertain to the water column and superadjacent airspace, and the Area, which is the ocean floor beyond the limits of national jurisdiction. On the high seas, certain traditional freedoms, including the freedom of fishing, are preserved in the Convention (Article 87). Subject to limited exceptions, states have exclusive responsibility for ships flying their flag (Article 92). Mineral exploration and exploitation activities in the Area are regulated by an international organisation known as the International Seabed Authority.

The Convention establishes general rules for most sectors of oceans use, including living and non living resource use, shipping and navigation, protection of the marine environment and marine scientific research. Those relevant to sustainable use of fisheries are considered in more detail below. Before doing so, it is worth commenting on some general features of the Convention that impact upon efforts to regulate resources, including the development of new rules.

Although the Convention purports to regulate all aspects of ocean use, some matters remain outside its remit, either because they could not be settled, such as military uses of the sea, or because states were simply unaware of the issues at the time of negotiation, such as the regulation of biodiversity. It is also important to note some of the Convention's limitations, because these may impede attempts to address issues *de novo*, or adapt existing rules to new circumstances. First, the Convention is a 'package deal'. It attempts to deal with all matters concerning ocean use in a single instrument. This means that many of its provisions are intrinsically connected, both legally and politically. This may limit attempts to adjust any single part of the Convention in isolation because other provisions were accepted as part of a *quid pro quo* during its negotiation. Secondly, the Convention's internal mechanisms for change, i.e. the amendment procedures contained in Articles 312–3, are procedurally difficult to initiate, requiring either the convening of a new conference or unanimity.[2] That said, it appears that some *de facto* development of the Convention's rules has been possible through other diplomatic processes, resulting in the adoption of two implementation agreements: the 1994 Implementation Agreement concerning deep seabed mining and the Fish Stocks Agreement of 1995.[3] Also, many of the Convention's provisions have been developed through global and regional instruments, many of which impact directly or indirectly on fishing activities. Finally, the Convention contains provisions for the settlement of disputes concerning its interpretation and application. However, the fisheries are largely excluded from their mandatory application meaning that there is reduced scope to develop the Convention's general fishing rules through case law.

As it may prove difficult to adapt some of the Convention's rules to new situations and changed circumstances, it is worth highlighting some provisions of the Convention that are crucial to the position of certain Commonwealth states. A number of the rights of land-locked states (Botswana, Lesotho, Malawi, Swaziland, Uganda and Zambia) are largely dependent upon the Convention. Although the Convention provides rights of transit to and from the sea in Article 125, this is contingent upon subsequent agreements, and it is thus essential that such states are able to negotiate effective access regimes with neighbouring coastal states. A useful model here is the Northern Corridor Transit Agreement, which provides Burundi, Rwanda, and Uganda access to the Kenyan port of Mombasa. Of course land-locked states may be in a poor bargaining position as regards transit passage, and so a considerable responsibility to cooperate falls upon neighbouring states. It is notable that several such transitory states include Commonwealth partners (Kenya, South Africa, and Tanzania) who will be responsible for proper facilitation of any transit rights. A second important right concerns access to marine resources. As land-locked states received little benefit from the expansion in exclusive coastal state control over ocean spaces and resources, some degree of benefit sharing was accorded to land-locked states through Article 70, which grants such states the rights to participate in surplus EEZ fisheries. However, such rights are contingent upon subsequent agreement and rather tenuous in practice.

Secondly, there is concern about the capacity of developing states to make submissions to the Commission on the Limits of the Continental Shelf. The initial time limit for such submissions, which was strictly framed in the Convention, required most states to make submission within 10 years of the entry into force of the Convention. Some degree of flexibility was advocated by Pacific Islands Forum States (Australia, Fiji, Marshall Islands, Micronesia (Federated States of), Nauru, New Zealand, Papua New Guinea, Samoa, Solomon Islands, Tonga and Vanuatu) due to delays in making the Commission operational, as well as some states' lack of financial and technical resources. As a result of this, for most states the deadline was set for May 2009. However, at a meeting of states parties to the Convention, it was agreed that states can make a preliminary submission, pending a full submission, and so allow for partial or undeveloped claims pending further preparation of the evidence.[4] For states such as Ghana, Guyana, India, Kenya and Malaysia, who have indicated an intention to submit, this provides important scope for developing their submissions.

## Domestic Fisheries

The main consequence of the EEZ was to bring around 90 per cent of commercial fishing under exclusive coastal state authority. Apart from securing the economic benefits of fishing to coastal states, the enclosure of oceans space was also intended to provide a basis for preventing the over-exploitation of fisheries, which were previously considered to be a common pool resource under the regime of the freedom of the high seas. Within the EEZ, coastal states enjoy certain exclusive rights to explore, exploit, conserve and manage living resources. These rights are subject to general conservation and utilisation duties, as set out in Articles 61 and 62 of the Convention. Thus coastal states must 'ensure through proper conservation and management measures that the maintenance of the living resources is not endangered by over-exploitation'. Such measures 'shall also be designed to maintain or restore populations of harvested species at levels which can produce the maximum sustainable yield, as qualified by relevant environmental and economic factors'. Coastal states are required to determine the total allowable catch (TAC) of fish that may be taken in its EEZ and, in order to 'promote the objective of optimum utilisation of the living resources' of its EEZ, are supposed to permit fishermen from other states to harvest that part of the allowable catch that cannot be taken by its own fishermen.

It should be clear from this brief restatement of coastal state conservation and management obligations that these operate at a high level of generality. Moreover, they predate the development of important management concepts, such as the precautionary principle and ecosystem approach, and it is not yet fully clear to what extent such approaches can be read into the Convention's conservation and management regime. Ultimately, the acid test of this will be the extent to which states advance such approaches in multilateral fisheries agreements and adopt relevant measures under domestic law. In defence of the Convention, it must be remarked that it was neither feasible nor intended that the Convention should provide comprehensive rules for the regulation of resources, rather it was expected that such rules would be developed within the Convention's basic framework. However, apart from straddling stocks and highly migratory species, the Convention provides no explicit mandate for the adoption of implementing agreements or regional fisheries agreements. This can be contrasted with the framework for the regulation of marine pollution, which explicitly mandates such developments. The result is that coastal states are left with a wide discretion to manage fisheries. This is not necessarily a problem if states take their responsibilities seriously, and it may provide coastal states with considerable scope for adapting fisheries regulations to suit local conservation and management imperatives.

However, the poor state of the world's fish stocks suggests that current conservation and management measures remain quite inadequate. According to FAO, approximately half of commercial stocks are fully exploited and producing catches at the maximum sustainable limit. A further 25 per cent of stocks are overexploited, significantly depleted or recovering from depletion (FAO, 2007). In part, this results from weaknesses inherent in the Convention, such as the use of the maximum sustainable yield (MSY). The MSY requires a degree of scientific data that is not always available or reliable, and may actually encourage the economic exploitation of stocks beyond truly sustainable levels. Moreover, the setting of catch limits is not free of political and socio-economic factors, and it is frequently the case that they are set above the levels recommended by scientists to meet the commercial and socio-economic needs of the fishing industry. Problems of science and decision-making aside, the position is compounded by legal failings, whereby states cannot readily be held accountable for their failure properly to conserve and manage their domestic fisheries (Barnes, 2006). In the first place, it may be difficult to prove they have breached their general conservation and management duties. Secondly, it is not clear that other states possess any legal standing, or indeed political will, to bring an international claim for the breach of conservation and management duties. Finally, although the Convention establishes a dispute settlement regime, as noted above, this explicitly excludes domestic fisheries from its compulsory scope.

## High Seas Fisheries

Significant commercial fisheries are located on the high seas. One adverse consequence of the emergence of the EEZ was that the exclusion of foreign distant water fishing fleets from the EEZ drove excess fishing capacity onto the high seas, thereby exacerbating the problems of over-exploitation of high seas stocks. On the high seas, states are free to fish subject to their other treaty obligations, certain rights, duties and interests of coastal states, and various duties set out in Articles 117–119 of the Convention. These Articles provide that states are to co-operate in the conservation and management of the living resources of the high seas, including through regional fisheries organisations where appropriate. Whatever form such co-operation takes, it should lead to the adoption of measures that are designed to maintain or restore populations of harvested species at levels that can produce the maximum sustainable yield, as qualified by relevant environmental and economic factors.

Many of the problems concerning domestic fisheries are replicated for discrete high seas fisheries. However, there are also problems specific to high seas fisheries. On a practical level, it is simply more difficult to police fishing activities occurring at vast distances from land. Again, such problems are compounded by shortcomings in the legal regime. Firstly, there is a failure of key fishing states to participate in existing multilateral treaty regimes pertaining to high seas fisheries. Secondly, such regimes have come under criticism for failing effectively to manage stocks falling under their remit. Thirdly, there remain gaps in the geographic and structural scope of fisheries management organisations. Thus certain areas of the high seas remain beyond the scope of cooperative arrangements, such as in the Indian Ocean and Southwest Atlantic. Elsewhere certain types of activity remain unregulated, such as the protection of marine habitats from destructive fishing techniques and the impacts of fishing on biodiversity.

Underlying these problems is a more fundamental weakness in the legal regime for high seas fisheries. Apart from in certain limited circumstances, such as piracy, exclusive jurisdiction to regulate activities on the high seas is reserved to the flag state. As a result the effectiveness of any fisheries regulation is contingent upon the capacity and willingness of the flag state to exert its authority over vessels flying its flag. Of course, it remains open for vessel owners to register their fishing vessels under a flag of convenience so as to avoid burdensome restrictions on fishing activities.[5] Evidence of increased illegal, unreported and unregulated (IUU) fishing activities on the high seas, point towards significant weaknesses in a system that depends upon exclusive flag state control and permits the easy registration of fishing fleets in states without any means or interest in controlling fishing activities. In the first place this requires measures to encourage more states to participate in the relevant treaties, and then to encourage these states to control vessels flying their flags. Unless states are willing and able to control their fishing fleets, then it may be that more radical steps will be required, such as permitting non-flag state enforcement of generally accepted fisheries regulations in cases of defaulting flag state enforcement.

## The Fish Stocks Agreement 1995

In the 1980s and early 1990s, disputes between coastal states and distant water fishing states revealed the limitations of the Convention's provisions on straddling and highly migratory fish stocks, which provide for only very general duties of cooperation. Ultimately, states' desires to avoid conflict resulted in the negotiation and adoption of the UN Fish Stocks Agreement. The Fish Stocks Agreement strengthened the Convention regime by requiring *inter alia*: the compatibility of coastal state and high seas management measures, the application of the precautionary approach, the strengthening of the role of regional fisheries management organisations or arrangements, articulating more clearly flag state responsibilities, and improving the mechanisms for enforcement of fisheries regulations. Initially hampered by a low level of ratifications, the Fish Stocks Agreement now has 75 states parties, including important distant water fishing nations like Japan and South Korea. The Agreement has had a marked impact on the practice of states and regional fisheries organisations, providing the impetus and framework for the establishment of a number of new regional fisheries organisations and the strengthening of existing organisations.

The importance of regional fisheries management organisations (RFMOs) in the management of high seas fisheries is recognised in both the Convention and the Fish Stocks Agreement, although the latter establishes a much more extensive and defined role for such organisations. In particular, the Fish Stocks Agreement seeks to make access to fish resources contingent upon participation in, or adherence with measures established by, the competent RFMO. However, treaty rules are only binding upon states parties to the agreement and any conservation and management requirements advanced in a regional fisheries instrument can be avoided by vessels flagging out to states that are not party to the Agreement. Although, as noted, the Fish Stocks Agreement has provided the basis upon which RFMOs have developed in recent years, there remains a number of constraints to the effectiveness of RFMOs and there have been calls to reform organisations in order, *inter alia*, to: improve their levels of performance and compliance with their measures; ensure their accountability; enhance their capacity to regulate fisheries; and widen participation in them (High Seas Task Force, 2006). The means to achieve these aims, however, are subject to considerable disagreement.

## The FAO Compliance Agreement

In 1993, the FAO adopted the Agreement to Promote Compliance with International Conservation and Management Measures by Fishing Vessels on the High Seas (Compliance Agreement). The two principal aims of the Compliance Agreement are to require the flag states of fishing vessels to exercise their responsibilities properly, and to require flag states to provide certain information to FAO about its vessels authorised to fish on the high seas, including action taken against its vessels engaging in activities that undermine the effectiveness of international conservation and management measures. Although the Compliance Agreement required only 25 ratifications in order to enter into force, it took nearly ten years to obtain that number, entering into force on 24 April 2003. As of 25 March 2009, it had 38 parties, including the EC, and the major high seas fishing states of Japan and South Korea, but not Russia. Much of the Agreement is aimed at flag of convenience states. However, only two such states, Belize and Cyprus (both from the Commonwealth), have become parties to the Agreement. Without much wider participation, the Agreement is unlikely to make much impact on high seas fisheries in general, or the free rider/flag of convenience problem in particular. Also, the Compliance Agreement is only effective in so far as substantive conservation and management measures exist, which is not the case with all high seas fisheries at present.

## The FAO Code of Conduct

The FAO adopted the Code of Conduct for Responsible Fisheries in 1995. It covers not only fisheries management in the EEZ and the high seas, but also matters such as aquaculture, post-harvest practices and trade. A non-binding instrument, the Code is intended to set out 'principles and international standards of behaviour for responsible practices with a view to ensuring the effective conservation, management and development of living aquatic resources, with due respect for the ecosystem and biodiversity'. Such guidance is available to states, intergovernmental agencies and others engaged in fisheries. Although the Code is not legally binding, parts of it are based on existing treaties, including the Law of the Sea Convention, the FAO Compliance Agreement and the Fish Stocks Agreement. As such, the Code is to be interpreted and applied in conformity with these instruments.[6] There is an expectation that states and others will apply and implement the Code, notwithstanding its voluntary nature, and that FAO will monitor such implementation.[7]

Implementation of the Code is supported through FAO education and training programmes, such as the FAO Interregional Programme of Assistance to Developing Countries and FishCode programme. Since its adoption, the Code has been developed through a series of non-binding International Plans of Action. To date, four such Plans have been adopted, dealing with the incidental catch of seabirds in longline fisheries, the conservation and management of sharks, the management of fishing capacity, and IUU fishing. The FAO has also adopted a number of technical guidelines to support implementation of the Code, as well as a Strategy for Improving Information on Status and Trends of Capture Fisheries within the framework of the Code.

## Illegal, Unreported and Unregulated Fishing (IUU)

The concept of IUU fishing has evolved in recent years to become one of the primary areas of concern for the international community. It is a significant concern for all Commonwealth states. Sub-Saharan Africa is known to be particularly affected by IUU fishing, but high incidences of IUU fishing have been recorded in the North and South Atlantic, Indian Ocean, Southern Pacific and Southern Ocean (High Seas Task Force, 2006). It affects many fisheries that Commonwealth states participate in or depend upon. A number of Commonwealth states (Australia, Canada, Namibia, New Zealand and the UK) have taken the initiative in responding to IUU fishing through the work of the High Seas Task Force and following up on its recommendations.

There is no single legal regime for IUU fishing, rather it is addressed through various instruments that seek to control a range of irresponsible and unsustainable fishing practices. In 2001, a non-binding International Plan of Action on IUU (IPOA-IUU) fishing was developed within the framework of the FAO Code of Conduct on Responsible Fisheries. It provides a 'toolbox' of measures to address the problem and covers all sectors of the fishing industry, from flagging of vessels to fishing authorisation to landing and market-based controls on fish trade. The IPOA-IUU depends upon national measures to implement it, and a number of further initiatives have been developed or commenced. However, concrete measures have been slow to emerge. For example, the EC has adopted a Community action plan to control IUU fishing in 2002, but has done little to implement it, pending the entry into force on 1 January 2010 of a Community Regulation on the issue.[8]

IUU fishing is generally perceived to be the product of overcapacity in the fishing industry, inadequate fisheries management and problems associated with either ineffective flag state control or the use of flags of convenience. Overcapacity is a problem usually associated with developed states, and often results in excess capacity being pushed out of domestic fisheries into international or third state fishing grounds where it exacerbates existing pressure on fisheries. Inadequate management is often attributed to developing states where there is a lack of experience and institutional capacity to manage large scale commercial fishing in a global fishing market. Some Commonwealth states (Belize, Cyprus, and St Vincent and the Grenadines) have been singled out for the use of flags of convenience in the fishing industry (Committee on Fisheries, 2001). Such states are often criticised for failing to ensuring that vessels flying their flag, and so subject to their exclusive jurisdiction on the high seas, adhere to international shipping regulations and sustainable fisheries standards. However, it is also clear that other states, whose nationals make ready use of such services, ought to have a role to play in tackling the problem. These are concerns facing all Commonwealth states.

Legal responses to IUU fishing can form only part of an overarching strategy to tackle the problem. Here regulation might usefully focus on defining specific offences, controlling access to fisheries, regulating port access, designing inspection regimes and catch certification schemes, and a system of penalties. In general the institutional control of IUU fishing will require some legislative measures. However, equally if not more important are measures to enhance the capacity of states to manage the problem through the implementation of effective fisheries management regimes. This requires aid, training, support and cooperation between states (Tsamenyi *et al.*, 2008).

## Final Comments

The Law of the Sea Convention is a significant achievement; one that Commonwealth states have played and continue to play an important role in developing. Widely ratified and wide-ranging, it establishes a framework for the regulation of most areas of ocean use. As a framework convention, it does not deal with every question of ocean use. This requires continued efforts to develop more detailed rules and provide answers to matters left unfinished by the Convention. Of course, such measures must be adopted in accordance with the general provisions of the Convention and in line with the political interests of its states parties.

The Convention regime for the conservation and management of resources of the EEZ or high seas has not been entirely successful. In general, this stems from poor management by states and, where appropriate, fisheries management organisations. Poor management results from a range of factors: a lack of knowledge about fisheries, marine ecosystems and human impacts; the failure to set catch limits strictly; excess fishing capacity; the use of inadequate and destructive fishing gear; and poor enforcement of fishing regulations. Some of these failures can be attributed to the Convention. For example, the Convention fails to articulate sufficiently precise duties to conserve and manage living resources of the high seas, or establish adequate institutional support for their regulation. Compliance remains problematic, with a strong reliance on flag state control, and the exclusion of fisheries from the compulsory dispute settlement provisions of the Convention. However, much responsibility also rests with states, a large proportion of which have failed to take responsibility for the proper and effective regulation of living resources. As such, measures to address the shortcomings of the Convention must do so in a way that is both acceptable to states and capable of being implemented in practice.

One can identify three broad goals for the future development of the law of the sea, goals which might be regarded as priorities for Commonwealth states. Firstly, the goal of universal participation in the Convention and related agreements remains a necessary step to establishing a consistent and certain basis for global and domestic fisheries regulation. This can best be pursued through diplomatic channels, and there is scope for steps at the Commonwealth level to encourage its members to participate fully in global and regional fisheries agreements. For example, Swaziland and Malawi are yet to ratify the 1982 Convention. No Commonwealth states in West Africa have ratified the Fish Stocks Agreement, and the level of ratifications by states in the Caribbean is a concern. In general the low number of states ratifying the Compliance Agreement is disappointing. The failure of states to ratify and implement key international agreements, when combined with the possible use and abuse of flags of convenience undermines the public regulation of ocean spaces. Thus, full participation by all interested Commonwealth states should be regarded as a priority and necessary step towards ensuring sustainable fisheries.

Secondly, formal participation in treaty regimes must be supported by measures to ensure the basic rules and principles are translated into practice. States and international agencies need to make continued efforts to ensure the effective implementation of the Convention and related instruments by states parties. This requires capacity building either directly through state sponsored mechanisms, or through the work of agencies such as the FAO or World Bank or Commonwealth Secretariat. In this context, there is a continued role to play for the Commonwealth Governance and Institutional Development Division. Finally, there is a need to ensure effective compliance with the Convention and related instruments. In line with this, the focus in recent years has moved away from the creation of new rules concerning international fisheries. Rather the emphasis is now on non-binding codes and governance measures designed to address the gap between fisheries regulation and fishing practice. The focus is not so much on law making. Of course translating international rules and best practice into domestic legal orders remains an important goal, but in many cases this requires an emphasis on the mechanisms of governance, and not simply formal rule making.

## Appendix I: Commonwealth Member States Participation in International Fisheries Agreements

| State | 1982 Convention *Ratification or accession* | Fish Stocks Agreement *Ratification or accession* | Compliance Agreement *Ratification or accession* |
|---|---|---|---|
| Antigua and Barbuda | Yes | | |
| Australia | Yes | Yes | Yes |
| Bahamas | Yes | Yes | |
| Bangladesh | Yes | Signatory only | |
| Barbados | Yes | Yes | Yes |
| Belize | Yes | Yes | Yes |
| Botswana | Yes | | |
| Brunei Darussalam | Yes | | |
| Cameroon | Yes | | |
| Canada | Yes | Yes | Yes |
| Cyprus | Yes | Yes | Yes |
| Dominica | Yes | | |
| Fiji | Yes | Yes | |
| Gambia | Yes | | |
| Ghana | Yes | | Yes |
| Grenada | Yes | | |
| Guyana | Yes | | |

| **State** | **1982 Convention** *Ratification or accession* | **Fish Stocks Agreement** *Ratification or accession* | **Compliance Agreement** *Ratification or accession* |
|---|---|---|---|
| India | Yes | Yes | |
| Jamaica | Yes | Signatory only | |
| Kenya | Yes | Yes | |
| Kiribati | Yes | Yes | |
| Lesotho | Yes | | |
| Malawi | Signatory only | | |
| Malaysia | Yes | | |
| Maldives | Yes | Yes | |
| Malta | Yes | Yes | |
| Mauritius | Yes | Yes | Yes |
| Mozambique | Yes | Yes | Yes |
| Namibia | Yes | Yes | Yes |
| Nauru | Yes | Yes | |
| New Zealand | Yes | Yes | Yes |
| Nigeria | Yes | | |
| Pakistan | Yes | Signatory only | |
| Papua New Guinea | Yes | Yes | |
| Samoa | Yes | Yes | |
| Seychelles | Yes | Yes | Yes |
| Sierra Leone | Yes | | |
| Singapore | Yes | | |
| Solomon Islands | Yes | Yes | |
| South Africa | Yes | Yes | |
| Sri Lanka | Yes | Yes | |
| St Kitts and Nevis | Yes | | Yes |
| St Lucia | Yes | Yes | Yes |
| St Vincent and the Grenadines | Yes | | |
| Swaziland | Signatory only | | |
| Tonga | Yes | Yes | |
| Trinidad and Tobago | Yes | Yes | |
| Tuvalu | Yes | Yes | |
| Uganda | Yes | Signatory only | |
| United Kingdom | Yes | Yes | EC accepted |
| United Republic of Tanzania | Yes | | Yes |
| Vanuatu | Yes | Signatory only | |
| Zambia | Yes | | |

## References

Barnes, R. (2006) 'The Convention on the Law of the Sea: An Effective Framework for Domestic Fisheries Conservation?' in Freestone, D., Barnes R. and Ong, D. (eds) *The Law of the Sea. Progress and Prospects*, pp. 233–60 (Oxford: OUP).

Committee on Fisheries (2001) *Report on the role of flags of convenience in the fisheries sector* (2000.2302(INI)).

FAO (2007) *State of World Fisheries and Aquaculture 2006* (Rome: FAO).

Freestone D. and Oude Elferink A.G. (2005) 'Flexibility and Innovation in the Law of the Sea – Will the LOS Convention amendment procedures ever be used?' in Oude Elferink, A.G. (ed.) *Stability and Change in the Law of the Sea: The Role of the LOS Convention*, pp. 169–221 (Leiden: Martinus Nijhoff).

High Seas Task Force (2006) *Closing the net. Stopping Illegal Fishing on the High Seas*. Governments of Australia, Canada, Chile, Namibia, New Zealand, and the United Kingdom, WWF, IUCN and the Earth Institute at Columbia University. Available online at *http://www.high-seas.org/*

Houston Lay, S., Churchill R.R. and Nordquist, M. (eds.) (1973) *New Directions in the Law of the Sea*, vol. I, p. 250 (Dobbs Ferry NY: Oceana Publications).

Tsamenyi, M., Palma, M.A., Milligan, B. and Mfodwo, M., (2008), *Development Impact of the Council Regulation Establishing a European Community System to Prevent, Deter and Eliminate Illegal, Unreported and Unregulated Fishing on Commonwealth ACP Member Countries* (Commonwealth Secretariat).

## Notes

1 1833 United Nations Treaty Series 396; (1982) 21 *International Legal Materials* 1261.

2 On the difficulties see Freestone and Oude Elferink (2005).

3 Agreement Relating to the Implementation of Part XI of the Convention of the United Nations Convention on the Law of the Sea 1994, 33 *International Legal Materials* 1309. Agreement for the Implementation of the United Nations Convention on the Law of the Sea of 10 December 1982 relating to the Conservation and Management of Straddling Fish Stocks and Highly Migratory Fish Stocks 1995, (1995) 34 *International Legal Materials* 1542.

4 Decision regarding the workload of the Commission on the Limits of the Continental Shelf and the ability of states, particularly developing states, to fulfil the requirements of article 4 of annex II to the United Nations Convention on the Law of the Sea, as well as the decision contained in SPLOS/72, paragraph (a). UN doc SPLOS/183 20 June 2008.

5 UN Document A/59/63 (2004), *Report of the Consultative Group on Flag State Implementation*.

6 Article 3.

7 Article 4.

8 Council Regulation (EC) No 1005/2008 of 29 September 2008.

# 5 Will You Be Eating Illegal Fish for Dinner Tonight?

Sandy Davies and Per Erik Bergh

## Introduction to Illegal Fishing

Will you be eating illegal fish for dinner tonight? The question may be surprising, but fish are becoming scarcer, and illegal fishing is on the increase. Governments have allowed too many fish to be caught, too many fish to be wasted, too many fish to be caught illegally, and too many people to get away with it. Moreover, such bad fishing methods destroy the places fish need to live and breed. This is without considering the additional damage arising from pollution, climate change, and competition for water.

More and more people want to eat fish for dinner, and this increasing demand for fish has created a 'supply gap', meaning that there is greater demand than there is supply to the global markets – a fact that has led to a thriving illegal market.

Illegal fishing both contributes to, and is a symptom of, the demise of fisheries and, unless action is taken, it is likely to spiral out of control. There is widespread international interest and cooperation in preventing this, and governments around the world are working together to tighten the controls on fishing vessels, fishing activities, the transport and trade of fish, in an effort to stop the spread of illegal fishing and to improve the management of their fisheries.

Nevertheless, illegal fishing is still taking place across every ocean of the world, resulting in lost income, jobs, fish for food, ecosystem productivity, and biodiversity, as well as damage to marine environments (FAO, 2008; SIF, 2008; World Bank, 2008). All of these outcomes take development targets, such as the Millennium Development Goals (MDGs), further out of reach – and inevitably the hardest hit are some of the world's poorest people who rely on fishing as part of their daily survival.

Illegal, unreported and unregulated (IUU) fishing, as referred to by fishery experts relates to specific types of fishing activity undertaken in defined locations by people or vessels fishing with particular laws or regulations applying to them.[1] However, when IUU fishing is being talked or written about by non-fishery experts it is often referred to as *illegal fishing*[2] – in this popular context it also includes a broad web of actions and entities including the illegal harvesting of fish and the shipment, processing, landing, sale, distribution and marketing of fish and fishery products.

Global estimates of illegal fishing (or IU fishing)[3] vary widely, falling between a value of $9 billion to $25 billion or between a catch of 10 and 28 million tonnes of fish (e.g. Agnew *et al.*, 2009 ($10 to 23.5 bn); Commission of the European Communities, 2007 ($15 bn); Pauly *et al.*, 2002 ($25 bn). The accuracy of these estimates is regularly debated by experts in the field, as are the methods used to calculate them, but few would argue that they offer a sense of the astounding volume of illegal fish harvested. If the lowest catch estimate is taken, this is about the same as the total annual catch for global inland capture fisheries,[4] and the value of the financial loss to the legitimate industry or fisher, is equal to at least 10 per cent of the value of legal global exports of fish and fishery products.[5]

This paper considers some of the progress that has been made towards stopping IUU fishing and hopes to encourage active engagement by all Commonwealth governments in the global fight against it.

## Progress in Tackling Illegal Fishing

The past decade has been the first to see substantial efforts to fight illegal fishing. The beginnings of an effective response now exist, and awareness has been raised of the destruction and negative impacts of IUU fishing. As we learn more, so the paradigm shifts; initially the focus was to target the illegal fishing vessels, while now, the emphasis has broadened to include the targeting of the illegal catch and to track it from 'hook to plate'. Initially, governments tended to drive actions against IUU fishing and these were mainly based on strengthening enforcement through monitoring, control and surveillance (MCS), while today, broader multi-stakeholder and regional approaches are becoming more common. Another significant change is that illegal fishing is no longer dealt with in isolation; it is becoming an integrated part of fisheries reform policies and approaches to the governance of fisheries. These shifts in approach are helping to move us closer to realistic and robust solutions, some of which are discussed in the next sections.

### International responses

The United Nations Food and Agriculture Organization (FAO) has taken a lead in promoting awareness and developing instruments to address illegal fishing, in particular through its Committee on Fisheries (COFI), which attracts a global audience and has initiated a number of worldwide responses. For example, in 2001 COFI adopted an International Plan of Action to Prevent, Deter, and Eliminate IUU Fishing (IPOA-IUU), in 2003 an IUU fishing resolution was adopted and in 2005 the Rome Declaration on IUU Fishing was produced.

Apart from these specific IUU focused instruments, the global legal framework also facilitates inter-governmental cooperation and the use of diplomatic pressure to stop IUU fishing through a range of UN driven fishery instruments,[6] some of which are legally binding while others are voluntary. At the COFI meeting in March 2009, delegations were able to provide reports on a range of issues, including how implementation of the National Plans of Action on IUU fishing (NPOA-IUU) was progressing; how negotiations are proceeding on the most recent anti-IUU instrument, the Port State Agreement; and on the merits of a global record of vessels as part of the global toolkit to overcome illegal fishing. The outcome of these interventions, although not yielding major new ideas apart from possibly the proposal for another set of international guidelines on how to manage the catching of unwanted fish that then end up being thrown away and wasted, did raise the noteworthy point that cooperative efforts in fighting IUU fishing are becoming increasingly valuable.

The priority issue was the FAO led negotiation towards a Port State Agreement (PSA), which earned the lion's share of the discussion and 5 of the 10 paragraphs in the IUU record. In brief, many so-called 'ports of convenience' are favoured by vessels engaged in IUU fishing, apparently with the authorities turning a blind eye in order to gain revenue from services and repairs, fish for processing and trade, and even income in the form of bribes. These ports, even if few, are the lifeline to illegal operators, as illegal fish must be landed somewhere (EJF, 2005; HSTF, 2006; SIF, 2008). The Spanish port of Las Palmas is an example, accepting many vessels listed on RFMO non-compliant or 'black' lists, and serving as an entry point for fish from West Africa into the EU. If successful, the PSA will sever these lifelines and close all ports to IUU vessels.

The Agreement is also about helping countries to improve information sharing systems and ability to undertake inspections, so that no port accepts, even unwittingly, foreign IUU vessels. COFI members are generally positive about the PSA, but there remain unresolved issues around the inclusion of all ports, the types of vessels that could be excluded, and any special requirements for developing countries. By the time this paper goes to press the PSA may be agreed, and such questions will be resolved, but the next challenge will be to implement it and to find mechanisms to help countries to close all ports to non-compliant operators. Only then could IUU be better controlled.

## National monitoring control and surveillance

Monitoring, control and surveillance (MCS) was once considered as the 'policing of fisheries' with the MCS team within a fishery administration ensuring laws were enforced and that those who broke them were charged. This remains essentially the case, but a contemporary MCS team goes beyond deterrence and detection of illegal fishing.

Today a balance needs to be struck between deterrence of illegal fishing and nurturing of voluntary compliance. MCS is no longer a question only for small scale, industrial or even foreign fishers who will also respond to pressure from their peers or diplomatic persuasion. Innovative approaches to voluntary compliance and to increase awareness are becoming more common – a recent example is the Worldwide Fund for Nature's International Smart Gear Competition for 2009. Over $50,000 of prize money is on offer for the best new fishing gears or fishing methods designed to reduce bycatch. The aim of this competition is to seek real-world fishing solutions that allow fishers to fish 'smarter' and more legally – by better targeting their intended catch and thus reducing potential illegal catches while also safeguarding birds, dolphins, sea turtles and other marine life from being unintentionally caught. This type of initiative is an example of how NGOs and civil society are leading the way in providing options for more balanced and inventive MCS systems.

Estimating compliance, or determining where, when and by whom illegal fishing is taking place remains a challenge for MCS. Progress is taking place, but practical and well accepted methodologies for accessing compliance are few. In Namibia, Bergh and Davies (2004) estimated the compliance levels for three important fisheries based on air and sea surveillance and observer reports over nine years. The results indicated the wide range in type and level of non-compliance across the different fisheries and the results guided the government in considering the different risk levels across the fisheries and allowed them to re-focus the MCS system. However, few countries have the type of data available over the time scales that Namibia does, making it difficult to apply this type of assessment. Compliance and associated risk estimates clearly require new approaches – it is time to bring together not just the modellers but also those working daily with fishers and vessels to take a fresh look at methodologies to better understand and quantify non-compliance in data poor fisheries. This will help administrations to better plan and target MCS activities.

As mentioned, there is an international plan of action (IPOA-IUU) that utilises a diverse set of measures to combat IUU fishing. Individual countries are required to extract the relevant and suitable measures from this international plan into their own bespoke national plan (NPOA-IUU). This gives national flexibility but maintains options for regional cooperation through coherent and comprehensive national building blocks, but most countries have fallen short of the FAO timelines for development, and of those countries that do have NPOA-IUU, few are implementing it on a day to day basis.

An NPOA needs to be central to a government's strategy to fight IUU fishing, but this is generally not happening. The process appears to be too complicated and too demanding of valuable resources. Instead the plans just end up becoming another plan alongside the strategic plan, the fishery management plan, the annual plan and the MCS operations plan. Prioritising and yes, planning, are essential as is training personnel, improving information systems, procuring and sharing MCS hardware, working with organisations such as the coastguard and navy and identifying the critical IUU risks. But are all these plans really making any difference on the ground – or are they just diverting attention and resources from the work at hand? This question needs to be asked and the answer needs to be taken seriously. Already over-stretched MCS administrations cannot just continue to be burdened without some serious consideration of the value of doing so.

## Regional and sub-regional strategies and approaches

Regional cooperation is needed, firstly across national waters, be they maritime or inland, secondly to manage the high seas and thirdly to find regional solutions to the migrant illegal fisher. High seas fisheries include fisheries located outside national waters or Exclusive Economic Zones (EEZs), as well as those that permanently straddle or move across these boundaries. Regional Fishery Management Organisations (RFMOs) are made up of the countries that have a fishing or coastal state interest in the fishery and that work together to co-ordinate and manage these high seas fisheries. Some RFMOs focus on particular species of fish (often tuna), while others focus on a geographical region.

All RFMOs have different remits with associated rules and regulations – these differences offer an opportunity for IUU vessels to find loopholes and gaps that allow them to carry on their business. But effort is being made to improve their ability to fight IUU fishing – an area which, with few exceptions, the member countries of RFMOs have rather neglected in the past.

The Lake Victoria Fisheries Organization (LVFO) is an example of an inland RFMO, that has recently focused attention towards IUU fishing and ways to combat it – the three countries around the lake are working together to protect the famous Nile Perch stock from the onslaught of uncontrolled fishing, that threatens the livelihood of the populations living around the Lake. Tanzania, a member of the LVFO responded to this in late 2008 by arresting hundreds of fishermen and confiscating their illegal fishing gear. The impact of this action has not been reported on, but it appears that the LVFO is encouraging different methods of both deterrence and voluntary compliance to improve the governance of the fishery and reduce illegal catches.

Advances in new technology also mark new options for controlling IUU fishing, and some of these benefit from regional approaches. For example, the Pacific Forum Fisheries Agency (FFA), which is concerned with the offshore tuna fisheries in the 17 independent Pacific Island countries plus Australia and New Zealand, is more able to afford advanced equipment than the individual countries. Recently the Agency reported that it is exploring the possibility of satellite monitoring to catch illegal tuna fishing vessels that are trans-shipping their catch at sea, away from port and the eyes of the law.

Dealing with the migrant illegal fisher, be it a small nomadic fishing dhow on one of the world's lakes, or a fishing vessel without any nationality but belonging to an international criminal syndicate, provides a problem that is best solved by cooperation. An example of a global attempt to tackle this is the work on the global record, an attempt to establish a global database of fishery related vessels. This approach raises many questions, but work continues and was endorsed by COFI 2009 as an initiative that the world community considers valuable and worthy of further investigation.

Within a regional context information exchange, coherent regulations and coordinated approaches, will make it harder for the culprits to break the law in one country and sail to another to offload and sell their illegal fish. Sharing of hardware, joint awareness campaigns, joint operations and shared capacity building and training programmes will also strengthen regional effectiveness.

## Trade measures and managing the chain of custody

In order to stop illegal fishing, it is not just those catching fish that must be stopped but also those that trade or transport it – catching, trading or transporting illegal fish must not pay. Illegal fish is often transferred at sea to a transport or factory vessel where it may be divided up for further transport, or processed into different product forms and then sent on its way to a range of brokers, importers and global markets by a range of transport methods.

Tracking illegal fish is a detective game – that requires cooperation between flag states, coastal states, port states, and market states, international and intergovernmental organisations, financial institutions, insurers and finally – the consumer. The consumer may be innocent – but so may many of the others that handle the fish along its journey before it ends up on the plate – or perhaps they are not – and that is where all good detective stories begin.

Unravelling the journey of illegal fish is an aspect of fighting IUU fishing that is gaining increasing interest. Trade and marketplace measures to stop the trade and handling of illegal fish include policies and practices that are used to track and regulate trade in seafood products such as catch monitoring, control measures, verification of product labelling and tracking through the chain of custody. The proliferation of these measures is adding increasing pressure on national MCS organisations to comply with ever stricter international standards. Although the need for such measures is evident, scrambling to implement them without first considering and researching the impacts and who will benefit will only add further complexities to an already daunting number of trade and market rules and regulations.

A recent example is that of the EU. In recognising its responsibility as a major fishery market, and as an importer of an estimated €1 billion of IUU fish annually, in 2008 the EU adopted a new comprehensive regulation aiming to control IUU fishing and to prevent IUU products from being sold on their market.[7] This regulation will enter into force on 1st of January 2010 and will include all vessels engaged in commercial fisheries that ultimately wish to sell the fish into the EU and also any EU national engaging in IUU activities. The regulation contains a range of provisions, the most critical one being the principle of traceability, by means of a catch certification scheme.[8] The EU claims, and many developed countries and NGOs have agreed that, if implemented fairly and fully, the regulation will serve to block IUU-caught fish from entering the EU market and is therefore a positive and decisive step.

However, developing countries can take a different perspective. In reflecting on a history of dealings with the EU including poor monitoring, control and reporting of their own distant water fishing fleet (DWFF), fishery access negotiations intertwined with political, trade and development interests, food safety measures (sanitary and phytosanitary (SPS) regulations) implemented and monitored in rather less than transparent ways, and then adding the very real concerns on quite how to cope with the additional work that will fall on them and the concern over how this may disrupt all the hard work that has gone into winning legitimate trade for their fisheries products – it is easy to see why the measures are less welcome.

## Solving illegal fishing through good governance

IUU fishing is now being seen more and more in light of overall fishery governance, and the promotion of good governance. Governance is seen as a range of processes whereby elements in society exercise power, authority and influence and put in place policies and decisions that impact on the economic and social development of society as a whole and everyone in it.

For IUU fishing this implies both looking at the wider impact and consequences of IUU fishing as well as assessing and involving as many of the processes and players as possible when seeking solutions. It also suggests that all the players, from fishers to governments, need to understand the benefits of playing by the rules and the consequences of not doing so. Understanding the impacts of illegal fishing, and more importantly the potential benefits of fishing legally and governing the fishery with fair and just regulatory systems, is important information that we need to know and that we need to share in media and promotion campaigns that get to the right audience. These campaigns are becoming more visual and impressive each year. NGOs have generally led the way with this type of work (e.g. EJF, 2007) and we are all slowly realising the need to be upbeat and modern in our marketing towards change.

Finding incentives to encourage good governance through accountability and simple clear systems to facilitate anti-corruption measures are also being discussed, and eyes are being opened to the fact that corruption in fisheries goes well beyond the simple definition of 'abuse of public office for personal gain'. A recent publication by Standing (2008) looks at corruption in industrial fishing in Africa and how illegal fishing by the distant water fishing fleet as part of an access agreement has often been treated leniently due to concerns of political interference, fear of damaging diplomatic relations or reducing development assistance, income from the agreements or paybacks to officials.

A regularly cited benchmark in fishery good governance action has been the International High Seas Task Force (HSTF). Formed following the Johannesburg World Summit in 2002 by a small group of countries, the HSTF wanted to fast track global action on illegal fishing. The players were fisheries ministers and international NGOs[9] who developed an action plan designed to combat IUU fishing on the high seas. The report of the Task Force and their plan called 'Closing the Net' (HSTF, 2006) outlines the key points of leverage that can brought to bear at national, regional and global levels in order to minimise the incentives to carry out IUU fishing on the high seas. The overall success of the HSTF is still difficult to judge; but it did promote at a high political level the need for good governance of fisheries and showed how any country or group of countries can lead by example and make innovative moves to stop illegal fishing.

## An Example for Change

Change is what we are striving for, and this change can be seen in many places. One example of how a dedicated effort, an inclusive approach and a focused objective can have an impact over a relatively short period is given in the Stop Illegal Fishing Programme. For a long time the Southern African Development Community (SADC) had talked about doing something new about illegal fishing – but it was only when Abraham Iyambo, the Minister of Fisheries and Marine Resources in Namibia, met with Gareth Thomas, a UK Minister for Development (DFID), that a new cooperation and momentum begun in 2006. The co-operation was based on the spirit of partnership between these two Ministers who were joined by others to create what became known as the Stop Illegal Fishing Programme (SIF).

The SIF Programme, over one year, facilitated many initiatives, including engaging partners through a website *www.stopillegalfishing.com*, outreach meetings and information flyers, building capacity through training and working groups, increasing knowledge by digging deeper into aspects of illegal fishing, increasing national capacity by working with countries to make NPOA-IUU useful and practical, and pulling this together in an exciting book that draws together issues of importance to illegal fishing in Southern Africa (SIF,

2008). These initiatives paved the way for practical pathways and concrete actions to stop illegal fishing and a new policy for SADC – the SADC Statement of Commitment to Stop Illegal Fishing. In July 2008 at a regional Conference in Namibia, the SADC Ministers responsible for marine fisheries cemented their political commitment against IUU fishing by signing the Statement and voicing their commitment to act on it.

The momentum gained from the Stop Illegal Fishing Programme and the Ministerial meeting has already led to enthusiasm to create a Ministerial Task Force for African Fisheries under the auspices of the African Union (AU) and NEPAD, that will again be supported by partnership with the UK government. The political commitment led to a range of actions against IUU fishing, including Namibia de-flagging the 'Paloma V' after it was implicated in illegal fishing in New Zealand in May 2008; Mozambique seizing the 'Antillas Reefer' in sovereign waters in July 2008; the Tanzania Fisheries Minister cancelling 69 licences held by foreign fishing vessels operating in the Indian Ocean, to control unregulated fishing in Tanzanian waters in July 2008; Madagascar arresting the Senegalese fishing vessel 'El Amine' for illegal fishing and for carrying 33 tons of processed shark products and 1.5 tons of lobsters on board in October 2008; and a joint sea surveillance patrol between South Africa, Mozambique and Tanzania leading to arrests of IUU vessels in both Mozambique and Tanzania in March 2009.

The Stop Illegal Fishing Programme received an overwhelmingly positive response from the governments, NGOs, fishing industry, civil society and programmes that it worked with. But it was the championing of the Stop Illegal Fishing campaign by the African leaders that provided the essential catalyst required to start the momentum and provide the political commitment for change. The message the ministers gave was clear – illegal fishing will not be tolerated nationally or regionally in Southern Africa and their action since has proved this commitment.

When the authorities that aim to stop this crime are more powerful and stronger than those committing it, then illegal fishing will be brought under control, and those guilty will be brought to justice. We all need to consider how we can contribute to this change so that we can continue to enjoy the fish on our plate!

## Useful web links

The Stop Illegal Fishing Programme: *http://www.stopillegalfishing.com*
The Marine Stewardship Council: *http://www.msc.org*
Environmental Justice Foundation: *http://www.ejfoundation.org*
Greenpeace: *http://www.greenpeace.org*
High Seas Task Force: *http://www.high-seas.org and http://www.closingthenet.info*
Illegal Fishing.info: *http://www.illegal-fishing.info*
Internet Guide to International Fisheries Law: *http://intfish.net*
One Fish: *http://www.onefish.org*
The International MCS network: *http://www.imcsnet.org*
EU Combating illegal fishing: *http://ec.europa.eu/fisheries/cfp/external_relations/illegal_fishing_en*

## References

Agnew, D.J., Pearce, J., Pramod, G., Peatman, T., Watson, R. *et al.* (2009) Estimating the worldwide extent of illegal fishing. *PLoS ONE* 4(2): e4570 doi: 10.137/journal.pone.0004570.

Bergh, P.E. and Davies, S.L. (2004) Against all odds; taking control of the Namibian fisheries, pp. 289–318. *Namibia's Fisheries: Ecological, Economic and Social Aspects*, Ed. Samaila, U.R., Boyer, D., Skogen, M.D. and Steinshamn, S.I.

Commission of the European Communities (2007) Communication from the commission to the European Parliament, the Council, the European Economic and Social Committee and the Committee of the Regions. *On a new strategy for the Community to prevent, deter and eliminate Illegal, Unreported and Unregulated fishing*. Brussels. COM(2007) 601 final.

EJF (2005). *Pirates and Profiteers: How Pirate Fishing Fleets are Robbing People and Oceans* (London: Environmental Justice Foundation). 24 pp.

EJF (2007) *Pirate Fish on Your Plate – Tracking illegally-caught fish from West Africa into the European market* (London: Environmental Justice Foundation).

FAO (2001) *International Plan of Action to Prevent, Deter and Eliminate Illegal, Unreported and Unregulated Fishing* (Rome: FAO).

FAO (2008) *The State of the World's Fisheries and Aquaculture* (Rome: FAO Fisheries and Aquaculture Department).

HSTF (High Seas Task Force) (2006) *Closing the net: Stopping illegal fishing on the high seas.* (Governments of Australia, Canada, Chile, Namibia, New Zealand, and the United Kingdom, WWF, IUCN and the Earth Institute at Columbia University).

MRAG (2008) *The Global Extent of Illegal Fishing.* (MRAG and Fisheries Ecosystems Restoration Research, Fisheries Centre, University of British Columbia).

Pauly, D.E. V. Christensen, S. Guénette, T. Pitcher, U. Sumaila, C. Walters, R. Watson and D. Zeller (2002) Towards sustainability in world fisheries, *Nature* 418, pp. 689–695.

SIF (2008) *Stop Illegal Fishing in Southern Africa* (Stop Illegal Fishing Programme. Stop Illegal Fishing, Gaborone, Botswana).

Standing, A. (2008) *Corruption and industrial fishing in Africa.* U4 Anti-Corruption Resource Centre, p. 29, U4 issue 2008:7 available on www.U4.no checked July 2009.

World Bank (2008) *The Sunken Billions – the economic justification for fisheries reform* (World Bank and FAO).

## Notes

1 This definition states that *illegal fishing* implies activities conducted in areas where laws apply (to a person or vessel) and they are broken, this may be in waters under national jurisdiction or seas that fall under areas managed by regional fishery management organisations (RFMO). It defines *unreported* fishing as relating to the non-reporting or misreporting of fishing activities, again in areas where laws or procedures apply to those fishing. Finally, *unregulated fishing*, perhaps the most challenging category, is defined as fishing by vessels that do not fly any nationality flag, or that fly the flag of a nationality that is not a party to the RFMO that governs where the vessel is fishing, or alternatively a vessel that is fishing in un-managed seas but breaks the obligations or rules of the flag that it flies.

This definition is adapted by the authors from the FAO IPOA-IUU definition of IUU fishing (FAO, 2001).

2 In this paper, unless otherwise stated, the authors use *IUU fishing* and *illegal fishing* interchangeably to describe the wider and more popular use of the phrases.

3 From a correct technical view IUU fishing often actually focuses more on illegal and unreported fishing rather than unregulated – this is called IU fishing.

4 FAO (2008) estimates the total inland capture fishery production in 2006 at 10.1 million tonnes.

5 *Ibid* – the 2006 total world export of fish and fishery products was $85.9 billion.

6 For example the 1982 UN Convention on the Law of the Sea (which is wider than just fisheries), the 1993 FAO Agreement to Promote Compliance with International Conservation and Management Measures by Fishing Vessels on the High Seas, the 1995 UN Fish Stock Agreement, and the 1995 FAO Code of Conduct for Responsible Fisheries.

7 Council Regulation establishing a Community system to prevent, deter and eliminate illegal, unreported and unregulated fishing, Council Regulation (EC) No 1005/2008 of 29 September 2008 *http://eur-lex.europa.eu/LexUriServ/LexUriServ.do?uri=OJ:L:2008:286:0001:0032:EN:PDF*

8 DG Fisheries website: Technical note on the catch certification scheme, *http://ec.europa.eu/fisheries/cfp/external_relations/illegal_fishing/pdf/technical_note01_en.pdf*

9 The Task Force includes fisheries ministers from Australia, Canada, Chile, Namibia, New Zealand and the UK, together with the Earth Institute, IUCN-World Conservation Union, WWF International and the Marine Stewardship Council.

# 6 Regional Fisheries Organisations and Sustainable Fisheries in the Caribbean

Milton O. Haughton

## Introduction and Background

The Caribbean Region's coastal and marine ecosystems and their biological diversity are valuable national and regional assets, which have been providing the countries with countless benefits in the form of food, employment, transportation, information, culture, and recreation. If these assets are used and managed well, they can make significantly enhanced and sustained contributions to a broad range of economic, social, cultural and nutritional goals.

The Caribbean Sea is a semi-enclosed sea with oceanographic features that are highly complex and variable both spatially and temporally. The dominant surface current system is the Caribbean Current, which enters the Caribbean basin in the southeast then flows west to northwest, exiting through the Yucatan Channel into the Gulf of Mexico. It then passes through the Florida Straits where it forms the Gulf Stream flowing northward along the east coast of the USA. There are also several east flowing counter currents, eddies, gyres, meanders and transient lenses of low salinity water from the major rivers, occurring on both sides of the main current system and along the coastal areas (Stevenson, 1981; Appeldoorn *et al.*, 1987).

At the political level there are 26 independent countries and 19 dependent territories. Sixteen of the independent countries and 18 of the 19 dependent territories are Small Island Developing States (SIDS). The countries range from among the largest and richest (USA), to among the smallest (e.g. Barbados, St Kitts and Nevis) and least developed (Haiti) in the world. There is significant disparity and diversity of national and regional governance and institutional arrangements (Mahon, 2001).

Caribbean countries have invested heavily in tourism, coastal aquaculture, marine fishing, maritime transportation infrastructure, housing, and basic sanitation infrastructure in the coastal zone in order to generate economic growth and development. These investments have not only brought about economic progress and prosperity, but have also resulted in profound changes in the natural environment and put the fragile coastal and marine ecosystems under severe stress.

Coastal communities and coastal and marine ecosystems in the Caribbean are threatened by several factors including rising sea temperatures and sea level associated with climate change, pollution, the environmental effects of tourism and agriculture, and infrastructure development and many other human activities. Concerns have also been increasing over the growing problem of over-fishing and the deteriorating status of fisheries resources and essential fish habitats. Protection, conservation and management of these coastal and marine resource systems are important, not only for their preservation and long-term viability but also because coastal communities are highly dependent upon them for livelihood and sustenance. This is particularly important in the Small Island Developing States, where majority of the population live in the coastal zone and derive their livelihood from the natural resources therein.

Fishing activities in the Caribbean, with few exceptions, are small-scale multi-species operations conducted from small open boats of up to 12 metres in length, using outboard motors. Small semi-industrial and industrial fisheries exist in Guyana, Suriname, Trinidad and Tobago and Belize for shrimp; The Bahamas and Jamaica for lobster and queen conch; and in Barbados and a few of the other eastern Caribbean Islands for pelagic species. A wide variety of fishing gears are used, including mesh-wire traps, longlines, handlines, gill nets, seine nets, trammel nets, and bottom trawls. The fisheries are generally open access. The industrial shrimp fisheries in Guyana and Suriname, and the industrial conch fisheries of Jamaica, which are managed by catch quota systems, are among the few exceptions.

Most of the traditional commercially important species and species groups are reported to be either fully developed or over-exploited. These include queen conch, spiny lobster, shrimp, shallow shelf reef-fishes (particularly members of the snapper (Lutjanidae) and grouper (Serranidae) families) and some of the highly migratory and straddling pelagic species which are managed by ICCAT (CRFM, 2005; 2006; 2007; FAO, 2008).

Frequent illegal incursions, unreported and unregulated (IUU) fishing by foreign vessels in waters under the sovereignty or jurisdiction of Caribbean states is a major problem, as is illegal fishing by local fishers operating within the territorial seas and EEZs. Local fishers frequently disregard regulations and rules regarding licensing and registration of fishers and fishing boats, gear limitations, closed seasons, closed areas, minimum size of fishes among other measures designed to protect and conserve the fishery resources (Haughton, 2003). Inadequate management is recognised as one of the main reasons for IUU fishing, over-fishing and degradation of essential habitats and ecosystems in the Caribbean.

Weak management and poor governance have been identified as major problems in many Caribbean countries (FAO, 2007). The capacity for effective fisheries management varies significantly between the different countries of the Caribbean region. While some countries have significant institutional and human capacities for fisheries management and conservation, in others fisheries are either managed in an elementary manner or unmanaged. The national fisheries departments which have been vested with the responsibility for fisheries management tend to be under-funded and under-staffed. According to Haughton (2003) even among the most developed countries in the Caribbean there are problems coping with the large number of species. From the region as a whole, there is little information on the status of the commercially important resources and even less on the hundreds of species of lesser importance to the region's fisheries.

An important characteristic of fish stocks within the Caribbean region and globally is that they are shared between two or more states and/or they extend beyond the maritime boundary of a state onto the high seas where they are exploited by other states and entities.

For this reason international and regional cooperation among states is a fundamental requirement for sustainable use, effective management and conservation of fisheries resources and their associated ecosystems. Regional Fisheries Bodies (RFBs) are key institutional mechanisms for governance and cooperation in addressing transboundary fisheries matters. There are also numerous other tangible benefits to regional cooperation in fisheries, especially among developing countries with limited financial, human and institutional resources for research, policy formulation and implementation, and monitoring and enforcement.

FAO provides a comprehensive definition, which says a regional fisheries body is:

> *... a mechanism through which three or more States or international organizations that are parties to any international fishery agreement or arrangement collaboratively engage each other in multilateral management of fisheries affairs related to transboundary, straddling and highly migratory fish stocks, through the collection and provision of scientific information and data, serving as a technical and policy forum, or taking decisions pertaining to the development and conservation, management and responsible utilization of the resources. It also refers to fisheries organizations that are charged with the production of scientific information to other bodies responsible for management. A RFB in other words is the instrument for fishery governance at the regional level. Such a body may be concerned with a single species or a group of closely related species or a whole range of species as well as the affiliated aspects of fisheries in a defined region or sub-region.* (Lugten, 1999)

Globally the legal and institutional framework for governance, sustainable use and conservation of marine fisheries and their ecosystems has evolved rapidly during the past 3–4 decades. This commenced with the development and codification of the law of the sea in the 1982 United Nations Law of the Sea Convention (the Montego Bay Convention). It lays down the rights and obligations of states and provides the legal basis upon which to pursue the sustainable development and protection of fisheries and the marine environment. The preamble speaks of a desire to establish:

> *... a legal order for the seas and oceans which will ... promote the peaceful uses of the seas and oceans, the equitable and efficient utilization of their resources, the conservation of their living resources, and the study, protection and preservation of the marine environment.*

Several provisions of the Montego Bay Convention speak to the issue of regional cooperation among states for the purpose of ensuring proper conservation and management of the living marine resources in areas under the jurisdiction of states (Articles 61, 63, 64 and 123) and on the high seas (Articles 118–119).

The Montego Bay Convention has been supplemented by a number of hard and soft law instruments which form an integral part of the global system for fisheries governance. Among the other significant instruments are the outcome of UNCED, in particular its Agenda 21; the 1995 Agreement on the Management of Straddling Fish Stocks and Highly Migratory Fish Stocks; The 1993 Compliance Agreement on High Seas Fishing; the 1995 FAO Code of Conduct for Responsible Fisheries, and its various International Plans of

Action; and finally the 2002 World Summit on Sustainable Development and the associated Johannesburg Plan of Implementation. These international instruments have provided further support for regional cooperation and the role of regional fisheries bodies in achieving sustainable use and conservation of marine fisheries.

## Regional Fisheries Bodies

In recent years, particularly since the crisis in global fisheries became evident in the 1980s, attention has been focused on mechanisms for improved governance of fisheries and ocean resources. In this connection, significant attention has been given to the role of regional fisheries bodies in resource management and conservation, and how to strengthen and empower them to be effective organisations. These concerns are reflected in various international fisheries and environmental agreements, both binding and non-binding, and in international forums.

There are presently about 46 RFBs globally, which on the basis of their mandate and function, can be classified either as a regional fisheries management organisation (RFMO), or a fishery advisory body. FAO (2001) defines RFMOs as '*intergovernmental fisheries organisations or arrangements, as appropriate, that have the competence to establish fishery conservation and management measures*'.

The key issue is that RFMOs have legal and regulatory power, and are required to adopt management and conservation measures which are legally binding on the states parties or within the area falling under their jurisdiction. FAO (2005–2009) identifies 20 RFMOs worldwide.

There are about 26 RFBs globally, which either do not have legal and regulatory power or in practice do not make management or conservation decisions or recommendations, which are binding on states parties. Their contribution to fisheries management and conservation should not, however, be overlooked as they coordinate research and stock assessment studies, build capacity of national researchers and stock assessment experts, and provide scientific and technical advice for their member states, which is then used for management decision-making and policy formulation.

### General duties and responsibilities of regional fisheries bodies

The general duties and responsibilities of regional fisheries bodies vary depending on whether the body is mandated to manage fisheries or is limited to providing advice and technical assistance to its members. RFBs in developing countries tend to have mandates aimed at assisting their members with fisheries development and capacity building as well as providing information on resource conservation and management. It is nevertheless possible to identify the common functions of regional fisheries bodies. These normally include:

- Provision of a forum for technical and policy discussion and decision-making in respect of fisheries development, conservation and management including coordinated, harmonised, concerted or joint programmes by their members; collection, storage, analysis and distribution of statistical data and information on fish catch and fishing effort in their area of jurisdiction;

- Coordination and conduct of scientific research and stock assessment studies to determine the state of fish stocks and marine ecosystems, management reference points such as maximum sustainable yields and total allowable catch in their area of jurisdiction;
- Provision of advice or recommendations for sustainable use, management and conservation of the fish stocks and marine ecosystems; which may include specific measures to be applied to achieve desired objectives;
- Representation or coordination of representation of their Members at international forums and negotiations involving fisheries.

Where the mandate of the organisation is fisheries management and conservation, or includes such responsibilities it may also be responsible for:

- Allocating fishing opportunities among its members which may include allocation of catch quotas;
- Establishing legally binding management and conservation measures such as size limits, area and seasonal closures of fishing grounds and conservation zones, limitation on fleet and fishing gear;
- Monitoring and enforcement of the management and conservation rules including the imposition of sanctions where the rules have been breached.

If on the other hand the organisation's mandate includes promoting development and sustainable use of the resources, its responsibilities will typically also include, *inter alia*:

- Developing human and institutional capacity of its members;
- Developing and maintaining relations with international and regional organisation and promoting technical cooperation;
- Technology transfer and information exchange on subject matters within its competence;
- Mobilising resources and managing development projects on behalf of its members.

## Regional Fisheries Bodies in the Caribbean

At present there are four regional fisheries bodies in the Caribbean whose mandates include facilitating international cooperation for the conservation and management of marine fish stocks and ecosystems. These are WECAFC, ICCAT, CRFM, OLDEPESCA and OSPESCA. A brief overview of the objectives and main functions and responsibilities of these organisations will be provided below. It should, however, be noted that there are other organisations involved in fisheries management and conservation in the Caribbean region that would not fall within the definition of a Regional Fisheries Body. These include, *inter alia*, the University of the West Indies; the Caribbean Community (CARICOM); the Organization of Eastern Caribbean States (OECS); the Association of Caribbean States (ACS); the United Nations Environment Programme (UNEP) Regional Coordinating Unit; and the Intergovernmental Oceanographic Commission, Sub-Commission for the Caribbean and Adjacent Regions (IOCARIBE).

## International Commission for the Conservation of Atlantic Tunas

The International Commission for the Conservation of Atlantic Tunas (ICCAT) is a regional fisheries body with responsibility for the management and conservation of tunas and tuna-like species in the Atlantic Ocean and adjacent seas (including the Caribbean Sea). The organisation was established in 1969 and has 47 member states including Barbados, Belize, St Vincent and the Grenadines, Trinidad and Tobago, the UK Overseas Territories in the Caribbean (Anguilla, Bermuda, British Virgin Islands, and the Turks and Caicos Islands) (ICCAT website). ICCAT is currently responsible for the management and conservation of about 30 tuna and tuna-like species.

The preamble of the Convention speaks of a 'resolve to conclude a Convention for the conservation of the resources of tuna and tuna-like fishes of the Atlantic Ocean' (ICCAT, 2007). The main objectives and responsibilities of ICCAT are:

- Collection, compilation, analysis and dissemination of fishery statistics;
- Coordinating research, including stock assessment, on behalf of its members;
- Developing scientific-based management advice;
- Providing a mechanism for contracting parties to agree on management measures, which may include catch quotas, minimum legal size; and closed season;
- Monitoring to ensure compliance with the management and conservation measures, including the imposition of sanctions where the rules have been breached; and
- Producing and disseminating publications on its activities.

An independent review of ICCAT was completed in 2008 to evaluate its performance as an RFMO and in fulfilling its mandate (Hurry *et al.*, 2008). The Report expressed serious concern about the state of some of the main species being managed by ICCAT, especially the eastern and Mediterranean bluefin tuna fishery, and to a lesser extent the yellowfin tuna, swordfish and albacore fisheries. It found that ICCAT member states were not taking their obligations seriously enough, especially in respect of the provision of accurate scientific data to the Commission, and compliance with agreed conservation and management regulations. The report concluded that ICCAT's performance could dramatically improve if the member states complied with their legally binding obligations. It also found that the Commission had many of the necessary tools in place to guarantee sustainable management and conservation of fisheries resources but they needed to be used properly and respected by all parties. The Report further concluded that the ICCAT Convention needed to be updated and modernised to take into account more recent developments in science and international law such as, *inter alia*: the ecosystem approach to fisheries; the precautionary approach; the special requirements of developing states; flag and port state controls; and systems for monitoring, control and surveillance and combating IUU fishing (Hurry *et al.*, 2008).

## FAO Western Central Atlantic Fisheries Commission

The Western Central Atlantic Fisheries Commission (WECAFC) is a Regional Fisheries Body established by FAO Council in 1973, to promote the effective conservation, management and development of the living marine resources of the area of competence of the Commission, in accordance with the FAO Code of Conduct for Responsible Fisheries, and to address common problems of fisheries management and development faced by Members of the Commission (FAO, 2006).

The Commission's main functions and responsibilities are, *inter alia*, to:

- Improve fisheries governance through institutional arrangements that encourage co-operation amongst members;
- Assist its members in implementing relevant international fisheries instruments, in particular the FAO Code of Conduct for Responsible Fisheries and its related International Plans of Action;
- Promote, coordinate and, as appropriate, undertake the collection, exchange and dissemination of statistical, biological, environmental and socio-economic data and other marine fishery information as well as its analysis or study;
- Promote, coordinate and, as appropriate, strengthen the development of institutional capacity and human resources, particularly through education, training and extension activities in the areas of competence of the Commission;
- Promote and facilitate harmonising of relevant national laws and regulations, and compatibility of conservation and management measures; and
- Assist Members with the conservation, management and development of transboundary and straddling stocks under their respective national jurisdictions.

The WECAFC is an advisory body promoting sustainable fisheries development and management in the Wider Caribbean region by providing both technical assistance to member states and a forum for discussion, information exchange and coordinating activities among the various sub-regional fisheries bodies such as CRFM and OSPESCA, and other states and entities active in fisheries within the Wider Caribbean region.

The Commission facilitates the coordination of research, education and training, and advises on policy directions for management of shared resources. To fulfil these objectives it has established the following subsidiary bodies: the Working Party on Assessment of

Marine Fishery Resources; the Working Party on Fishery Economics and Planning; the Committee for Development and Management of Fisheries in the Lesser Antilles; and a Scientific Advisory Group. The advice and information produced by WECAFC are documented and disseminated to Members of the Commission. Being established in accordance with the FAO Constitution, the WECAFC structure receives additional technical support and expertise for its work from the central FAO structure that serves similar needs in fisheries at the global level (Singh-Renton, 2005).

## Latin American Organization for Fisheries Development

The Latin American Organization for Fisheries Development (OLDEPESCA) is an RFB that was established in 1982 to promote food security by development and use of fishery resources of Latin America, by strengthening regional cooperation among the states parties. OLDEPESCA's scope of work includes the marine waters under the jurisdiction of its members in the Pacific and Atlantic Oceans. There are 13 member states, including two Caribbean Commonwealth states, Belize and Guyana (OLDEPESCA, 2000; OLDEPESCA, 2003–2007).

According to Article 4 of the Agreement establishing OLDEPESCA:

> *[the] main purpose of the Organization is to meet Latin American food requirements adequately, making use of Latin American fishery resource potential for the benefit of Latin American peoples, by concerted action in promoting the constant development of the countries and the permanent strengthening of regional cooperation in this sector.*

The main functions and responsibilities of the Organization are *inter alia* to:

- Promote the utilisation of fishery resources, and preserve the marine and freshwater environments through the application of rational policies for the conservation of the resources;
- Encourage and strengthen regional cooperation in the development and rational exploitation of marine and freshwater fishery resources;
- Enhance food security by increased supply of fish of sufficient nutritional value, at prices, preparation and presentation which are in keeping with the needs of low-income inhabitants of the region;
- Promote marketing systems to expand regional exchange of fishery products;
- Improve and strengthen the productive, institutional, organisational and human resources capacity of the sector; and
- Promote the use of joint negotiating capacity for the Latin American region, as well as to mobilise international technical and financial cooperation, through concrete regional coordination and cooperation in fisheries.

## Organization for Fisheries and Aquaculture of the Central American Isthmus

The Organization for Fisheries and Aquaculture of the Central American Isthmus (OSPESCA) is an RFB established in 1995 by the Central American Countries, including Belize, a Commonwealth Country. OSPESCA's main purpose is to enhance sustainable development and management of fisheries and aquaculture by coordinated regional action, thereby strengthening the Central American integration process.

The main functions of the OSPESCA are, *inter alia*, to:

- Implement strategies to realise objectives of the common fisheries policies for Central America;
- Promote and implement the Regional Framework Treaty on Fisheries and Aquaculture;
- Coordinate inter-institutional and inter-sectoral efforts that are regional in scope and apply the ecosystem approach to fisheries;
- Coordinate efforts to harmonise and implement fisheries and aquaculture legislations of its members;
- Formulate and implement strategies, programs, projects, or regional agreements for fisheries and aquaculture;
- Coordinate regional participation in international forums related to the fisheries and aquaculture.

## Caribbean Regional Fisheries Mechanism

Arising from UNCLOS, the member states of the Caribbean Community have recognised the need for regional cooperation to ensure sustainable development and management of their fisheries and have undertaken a number of projects in this direction. In 1991, with financial and technical support from the Government of Canada, CARICOM countries launched the ten year, $25 million (Canadian) CARICOM Fisheries Resource Assessment and Management Programme (CFRAMP). The overall goal of CFRAMP was to promote sustainable use and conservation of the fisheries resources of CARICOM member states. The specific purpose of the project was to enhance the institutional capacity and information available for fisheries management planning and policy formulation. The main activities of CFRAMP included among other things, the establishment of a permanent regional fisheries body which would continue to support and coordinate sustainable fisheries development, management and conservation after CFRAMP ended (CRFAMP, 1991).

CFRAMP was the first major fisheries project of its type, focused as it was on adopting a regional approach to sustainable development and management of the fisheries resources of the CARICOM states. When the Project came to an end in 2002, the Heads of Government signed the Inter-governmental Agreement establishing the Caribbean Regional Fisheries Mechanism (CRFM).

The CRFM is an independent regional fisheries body established to promote sustainable use of the living marine and other aquatic resources by the development, efficient management and conservation of such resources. The objectives of the CRFM as stated in Article 4 of the Agreement establishing the Organization (CRFM, 2002) are:

- Efficient management and sustainable development of marine and other aquatic resources within the jurisdictions of member states;
- Promotion and establishment of co-operative arrangements among interested states for the efficient management of shared, straddling or highly migratory marine and other aquatic resources;
- Provision of technical advisory and consultative services to fisheries divisions of member states in the development, management and conservation of their marine and other aquatic resources.

The main functions and responsibilities of the CRFM are, *inter alia*:

- Promotion of efficient management, conservation and development of shared, straddling and highly migratory marine and other aquatic resources of the Caribbean Region through attainment of competence over the resources and through co-operation with competent organisations as the case may be;
- Development and maintenance of relations with national, sub-regional and regional institutions and bodies and international institutions and organisations the work of which have an impact on the fisheries within the Region;
- Promotion of human resource training and development in the fisheries sub-sector at the professional, technical and vocational levels in member states;
- Promotion and support for programmes designed to establish, facilitate and strengthen fisheries research, including the acquisition and sharing of relevant data in member states;
- Promotion of technical co-operation in the fisheries sub-sector, including technology transfer, information exchange and networking among states of the Caribbean Region and beyond;
- Encouragement of cooperation among the member states in order to avoid disputes or to resolve them in a peaceful manner;
- Promotion of the protection and rehabilitation of fisheries habitats and the environment generally;
- Encouragement of the establishment of effective mechanisms for monitoring, control and surveillance of fisheries exploitation;
- Development of arrangements for sustainable fisheries management and development

in member states based upon the best available technical or scientific data and information;

- Examination and consideration of action taken by member states and third states which may prejudice arrangements for sustainable fisheries management and development; and
- To receive reports on new arrangements made between member states and third states with respect to the conservation and management of fisheries.

CRFM is made up of three bodies as follows:

- A Ministerial Council (Ministers Responsible for Fisheries of member states) that has primary responsibility for, *inter alia*, determining the policies of the Organisation, resource allocation, cooperative agreements, and related decision-making;
- The Caribbean Fisheries Forum (heads of national fisheries administrations) which provide technical leadership to the Organisation, including the provision of scientific advice to the Ministerial Council, and oversight to the operations of the CRFM Secretariat; and
- The CRFM Secretariat, which is the technical unit responsible for day-to-day coordination and execution of the work programmes; collaborating with national fisheries authorities; mobilising resources; and managing the institutional networking to promote its optimal involvement and efficient functioning.

There are presently 17 member states of the CRFM. Membership is open to the CARICOM members and associate members. These are: Anguilla, Antigua and Barbuda, Barbados, Belize, Dominica, Grenada, Guyana, Haiti, Jamaica, Montserrat, St Kitts and Nevis, St Lucia, St Vincent and the Grenadines, Suriname, The Bahamas, Trinidad and Tobago, and Turks and Caicos Islands.

The CRFM has established a number of fisheries resource working groups which have met annually under the umbrella of the CRFM Annual Scientific Meetings. They compile and analyse available statistics and conduct assessment studies to determine the state of various fish stocks and provide management advice to the member states. The working groups include: the Large Pelagic Fisheries Resource Working Group; the CRFM Small Coastal Pelagic Fisheries Resource Working Group; the Reef and Slope Fisheries Resources Working Group; the Conch and Lobster Resource Working Group; and the Shrimp and Groundfish Working Group (CRFM, 2003).

The reports arising from the CRFM Scientific Meetings are documented and distributed to member states and other stakeholders by the CRFM Secretariat in the CRFM Fishery Report series (CRFM, 2005; CRFM, 2006; CRFM 2007). The results and recommendations are also formally submitted first to the CRFM Forum for consideration and then, with their endorsement, to the Ministerial Council for approval or information.

The CRFM contributes significantly to all aspects of fisheries management planning and decision-making in its member states, including data collection, analysis and data management, research, dissemination of scientific and technical information, preparation of national fisheries management plans, and strengthening national capacity for management. The CRFM also contributes significantly to the efforts of ICCAT in the Caribbean by coordinating data collection and research on the tunas and tuna-like species of interest to ICCAT, and facilitating information exchange and reporting and general interaction between CRFM member states and ICCAT. The CRFM Secretariat participates in the annual scientific and Commission Meetings of ICCAT providing technical support to its member states present and represents the interest of those that are not present.

At the heart of the CRFM Agreement is a commitment to sustainable use by conservation, protection and efficient management of the fishery resources of the Caribbean through closer and deeper cooperation among the member states in accordance with the principles laid down by the 1982 United Nations Convention on the Law of the Sea and associated agreements, including the FAO Code of Conduct for Responsible Fisheries and its various International Plans of Action. It is also a bold attempt to be more self-reliant and to take greater control over the aquatic resources on which so many of people and communities depend for their livelihood and sustenance.

Notwithstanding the above, for present purposes, the question to be asked after seven years of operations, is whether or to what extent the CRFM is achieving sustainable management of the shared fisheries resources within the jurisdiction of its member states as provided for in Article 4 of its constitution. The simple answer is that these matters have not been the main concern of the CRFM. Although the 2002 CRFM Agreement empowers the Organisation to function as a RFMO, the members have neither sought to develop nor exercised this power to date. Although the CRFM may not have power to make decisions that are directly binding on member states, the Ministerial Council can make conservation and management recommendations that member states would be obliged to implement. Member states prefer instead to use the CRFM as an advisory body providing scientific information, recommendations and other forms of support to the Member States, to allow them to make informed decisions and strengthen their capacity to govern and manage the fisheries themselves. Member states are more comfortable with this approach and have resisted attempts to move in the direction of giving greater attention to addressing resource management and conservation issues, and actively implementing regional management schemes. Whether or not the CRFM functions in this capacity, i.e. as a regional fisheries management organisation, is ultimately a question of political will and the priority attached to resource management and conservation by its Member states. This must be so because decisions to manage and regulate should not be based solely on science because they involve judgment of acceptable versus unacceptable risks, as well as assessments of acceptable versus unacceptable costs of regulation and management (Ellis, 2006).

As McDorman (2005) explains, the doctrine of state sovereignty is the core reality in respect of decision-making within RFBs, and thus the need for and influence of political will. The consequence of the doctrine is that, except in rare circumstances, decisions and obligations are not binding on a state without the direct consent of that state. 'In the RFMO decision-making context, the centrality of state sovereignty is clear: (i) whether a specific decision is made within an RFMO is in the hands of the states that are members of the RFMO; and (ii) the creation or amendment of the procedure or mode of decision-making is in the hands of states since international treaties negotiated by states establish RFMOs' (McDorman, 2005).

However, as the commercially important fish stocks and the coastal and marine ecosystems continue to decline; as pressure increases, both internally and externally, for greater food security and sustainable social and economic benefits; the need for good governance and effective resource management and conservation is becoming more urgent. There may be no credible alternative to strengthening the capacity and performance of the CRFM or another regional fisheries body such as WECAFC in order that fish stocks may be better conserved and managed and thus able to make sustainable contributions to the Region's economic development and food security. With time member states may decide to attach greater weight to the principles of sustainability and conservation of the marine resource systems.

There are reasons to be optimistic about the future of fisheries in the CARICOM member states. This is because in 2003 the Conference of Heads of Government of CARICOM mandated the preparation of a Common Fisheries Policy and Regime (CFP&R) at the CARICOM level. The basic objective of the CFP&R is to enhance the contribution fisheries make to food security, and social and economic development by providing a harmonised comprehensive framework for responsible and sustainable exploitation and conservation of the fisheries resources in the Region. Although the agreement establishing the policy is still being negotiated, a few useful observations may be made about the progress to date. Firstly, it appears that member states are prepared to give higher priority to conservation and sustainable use of the marine resources. On the other hand one of the main obstacles to the conclusion of the negotiations is centered on the question of how much power should be vested in the implementing body (the CRFM) in respect of fisheries conservation and management. This matter has still not been resolved. It should be noted that the initiative to establish a common policy is being pursued within the context of a broader commitment by the CARICOM countries to closer regional cooperation for economic and social development through the establishment of a CARICOM Single Market and Economy, which is already well underway (CARICOM, 2002).

## Conclusion

The RFBs in the Caribbean were established to promote regional cooperation in the development, conservation and management of the fisheries of the Region, in accordance with relevant principles of international law. As noted above most commercially important marine species in the Caribbean are either shared, straddling or highly migratory species which can only be effectively managed through cooperative efforts of the states and entities involved in the fisheries. For present purposes, the main strength of the RFBs in the Region is that they provide a framework for managing shared stocks throughout their range of distribution, taking into account the need for harmonised or common measures across national jurisdiction and on the High Seas.

With the exception of ICCAT, the RFBs in the Caribbean were not established exclusively for the purpose of conservation and management of fish stocks. Their primary role is rather to provide advice to their Members to develop and manage their fisheries through cooperation in data collection, research, stock assessment, the provision of technical assistance, resource mobilization and other similar interventions. It is also clear that the RFBs are able to carry out these responsibilities much more cost effectively than if each country were to attempt them individually. Sustainable use and effective management and conservation of the shared, migratory and straddling stocks in the region, especially those which are not currently being managed by ICCAT, remain a serious concern and challenge.

However, the role of RFBs with an advisory mandate and the contributions they can make to national governance and the work of RFMOs should not be underestimated. Their activities may lead to improved national fisheries governance and harmonised regional measures for conservation and management.

Notwithstanding the above, much more needs to be done to improve governance frameworks and management of fisheries in the region. The mandates of the RFBs such as CRFM and OSPESCA appear to be broad enough for these organizations to play more significant roles in decision-making related to resource conservation and management, but only if their Members so desire. Whether or not these bodies function in this capacity (as an RFMO) seems to be a question of political will and the priority attached to resource management and conservation by the Members.

Regional fisheries bodies, including RFMOs, are not supranational organisations. They are made up of sovereign states that come together for the purpose of addressing their common interests and desire to obtain sustainable benefits or other objectives from their fisheries. Therefore, the power and effectiveness of RFBs are determined by the will of their members.

What is clear is that as coastal and marine ecosystems and their fish stocks continue to decline the need for good governance, and effective resource management and conservation is becoming more urgent in the Caribbean. Strengthening existing RFBs and their performances in order that fish stocks may be better conserved and managed is arguably the only realistic and cost effective option available to the countries at this time. Caribbean states therefore need to individually and collectively review and clarify their national and regional priorities for the fisheries sector and then take the steps necessary to achieve their desired goals.

Recent advances in science, technology and international law support a greater role for RFBs in regional fisheries governance and decision-making to achieve sustainable fisheries. Further reforms and strengthening of the mandates and institutional capabilities of RFBs in the Caribbean should be undertaken to address existing weaknesses and emerging issues such as the application of the precautionary approach and the ecosystem approach both regionally and at the national level; improving systems for collection, analysis and provision of good quality, reliable scientific information for decision making and policy formulation; strengthening decision-making procedures to facilitate the adoption of conservation and management measures before stocks reach and intensifying the fight against IUU fishing.

Whether or not Caribbean states will decide to strengthen RFBs and use them as a vehicle for regional management and conservation of marine fisheries is ultimately a question of political will.

## References


Appeldoorn, R. Dennis, D.G. and L. Monterossa (1987) *Review of the shared demersal resources of Puerto Rico and the Lesser Antilles Region*. FAO Fish Report. 383, pp. 36–106 (Rome: FAO).

CARICOM (2002) *Revised Treaty of Chaguaramas Establishing the Caribbean Community Including the CARICOM Single Market and Economy* (Georgetown: Caribbean Community Secretariat).

CFRAMP (1991) Memorandum of Understanding between the Caribbean Community and the International Centre for Ocean development (Belize City: CFRAMP).

CRFM (2002) *Agreement establishing the Caribbean Regional Mechanism* (Belize City, Belize: CRFM).

CRFM (2003) *Report of the First Meeting of the Caribbean Fisheries Forum of the Caribbean Regional Fisheries Mechanism*. CRFM: Belize. 25 pp.

CRFM (2005) Report of the First Annual CRFM Scientific Meeting. CRFM Fishery Report No. 11 (Belize City: CRFM).

CRFM (2006) *Report of the Second Annual CRFM Scientific Meeting, Port of Spain, Trinidad and Tobago*, 13–22 March 2006. CRFM Fishery Report 2006. Vol. 1.

CRFM (2007) *Report of the Third Annual CRFM Scientific Meeting, St Vincent and the Grenadines*, 17–26 July, 2007. CRFM Fishery Report 2007. Vol. 1.

Ellis, J. (2006) Overexploitation of a valuable resource? New literature on the precautionary principle. *European Journal of International Law*, 17(2), pp. 445–462.

FAO (2001) *International plan of action to prevent, deter and eliminate illegal, unreported and unregulated fishing*. (Rome, FAO).FAO (2005–2009) Regional Fishery Bodies (RFB) – Web Site. Strengthening regional fisheries governance. Text by David Doulman. In: FAO Fisheries and Aquaculture Department [online]. Rome. Updated 27 May 2005. *http://www.fao.org/fishery/topic/14908/en* [Accessed 19 March 2009].

FAO (2006) *Revised Statutes of the Western Central Atlantic Fishery Commission* (WECAFC). FAO Council Resolution 1/131 of November 2006 (Rome: FAO).

FAO (2007) *Regional Workshop of factors contributing to Over-exploitation and Unsustainability in fisheries in CARICOM Countries*, Barbados, 6–8 December 2006.

FAO (2008) *Status of the stocks of the Western Central Atlantic*. FAO Statistical Area 31. (WECAFC/XIII/08/02E) WECAFC 13th Session and WECAFC Lesser Antilles Fisheries Committee 10th Session, Cartagena de Indias, Colombia, 21–24 October 2008.

FAO Project GCP/INT/788/JPN.

Haughton, M (2003) *Compliance and Enforcement of Fisheries Regulations in the Caribbean*. GCFI 54, pp. 188–2001.

Hurry, G.D., Hayashi, M. and Maguire, J.J. (2008) *Report of the Independent Review, International Commission for the Conservation of Atlantic Tunas (ICCAT)*. September 2008. PLE-106/2008. ICCAT (Madrid).

ICCAT (2007) Basic Texts, 5th Revision. International Commission for the Conservation of Atlantic Tunas, Madrid, Spain. 117 pp.

ICCAT (2009) Contracting Parting. [Cited 20 March 2009]. *http://www.iccat.int/en/contracting.htm*

Lugten, G.L. (1999) A Review of measures taken by regional marine fishery bodies to address contemporary fishery issues. *FAO Fisheries Circular*. No. 940 (Rome, FAO).

Mahon, R (2001). Concept paper for project on sustainable management of the shared living marine resources of the Caribbean Sea Large Marine Ecosystem (CLME) and Adjacent Regions. Caribbean Large Marine Ecosystem Project (IOCARIBE).

McDorman, T.L (2005) Implementing existing tools: turning words into actions – decision-making processes of regional fisheries management organizations (RFMOs). *International Journal of Marine and Coastal Law*, 20(3/4), pp. 423–457.

OLDEPESCA. *Constitutional Agreement of the Latin American Organisation for Fisheries Development. Que es OLDEPESCA?* (Lima).

OLDEPESCA (2003–2007) Paises miembros Organización Latinoamericana de Desarrollo Pesquero [Accessed 20 March 2000] *http://www.oldepesca.org/Dnnportal/*

Singh-Renton (2005) A. *An overview of a regional fishery body, with emphasis on the type serving as a regional fishery management organization*; and B. *A preliminary examination of the issues pertaining to establishment of a regional fishery management organization within the Caribbean region*. Discussion Paper (Revised 8 April 2005). Report of the Third Meeting of the CRFM Working Group on the Common Fisheries Regime (Belize: CRFM Secretariat).

Stevenson, D. K (1981) A *review of the marine resources of the WECAFC Region*. FAO Tech Pap 211. FAO, Rome, 134 pp.

# 7 Developments in Cooperative Approaches to Fisheries Management in the Pacific Islands Region

Quentin Hanich and Martin Tsamenyi

## Introduction

The Pacific islands have developed a collaborative approach to fisheries management that has set global precedents in regional fisheries cooperation and significantly boosted their capacity to manage the region's tuna fisheries and progress their national interests. Beginning in 1947 with the founding of the Secretariat of the Pacific Community, the region has established various scientific, technical and policy forums and negotiated a number of sophisticated regional fisheries agreements.

Regional co-operation is vitally important in the Pacific islands region due to the migratory nature of the tuna fisheries and the limited capacity of most Pacific island states[1] to take advantage of their rights and discharge their obligations under the United Nations Law of the Sea Convention (LOSC) following the extension of maritime zones of jurisdiction. In response, Pacific island states have collectively established some of the world's most sophisticated and advanced cooperative tools. Similarly, through cooperation, the Pacific island states have developed a collective influence in fisheries negotiations that is arguably far greater and more effective than what they could achieve individually. These regional achievements are particularly impressive given the limited capacity of the region, and the economic and governance challenges that many of the islands face.

## The Pacific Island States and Territories

The term South Pacific is usually used to describe the island states and independent territories in the Western and Central Pacific Ocean (WCPO). Geographically, the region extends from French Polynesia in the east to Papua New Guinea in the west, with a number of unique characteristics.

The region is vast, encompassing a unique grouping of small island states that includes some of the world's smallest countries. The combined Exclusive Economic Zones (EEZs) of the Pacific island states cover roughly 30,569,000 km$^2$ of the Western and Central Pacific Ocean (WCPO)[2] and include some of its most productive waters. Yet the combined landmass of these island states is only 552,789 km$^2$, of which 84 per cent is found in Papua New Guinea (Gillett, 2005).

The island states and territories in the region are at different stages of political and economic development. Twelve are politically independent but with substantial dependence on foreign aid (Federated States of Micronesia, Fiji, Kiribati, Marshall Islands, Nauru, Palau, Papua New Guinea, Samoa, Solomon Islands, Tonga, Tuvalu and Vanuatu). Two are independent but affiliated with New Zealand (Cook Islands, Niue). One is part of New Zealand (Tokelau) and the rest are dependent territories of either the United States, France, or the United Kingdom. (American Samoa; Guam; Northern Marianas; French Polynesia; New Caledonia; Wallis and Futuna; and Pitcairn Islands). Nine Pacific island states are Commonwealth members: Fiji (membership suspended in December 2006 following the military coup), Kiribati, Nauru, Papua New Guinea, Samoa, Solomon Islands, Tonga, Tuvalu and Vanuatu.

For the majority of Pacific island states, fisheries provide the only economic activity. It is not surprising, therefore, that most domestic and international issues in the region revolve around the marine environment.

**Figure 1: WCPF Convention Area**

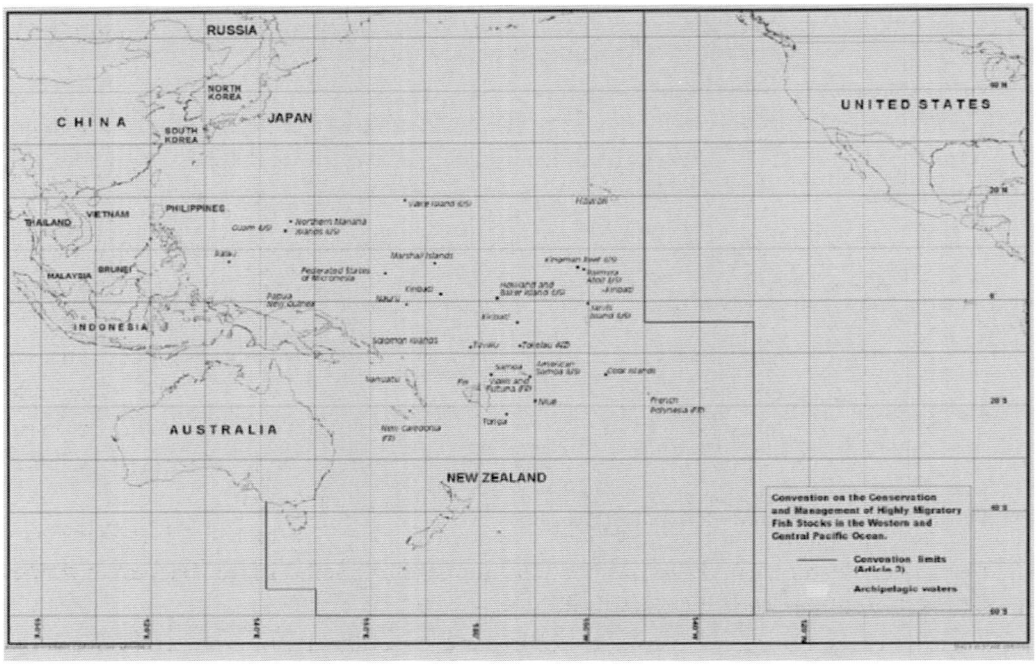


Source of map: *http://www.wcpfc.int/*

## Fisheries Governance Challenges

Many Pacific island states are in a precarious condition (ForSEC, 2005a) with low economic growth, political instability and significant weaknesses in their institutions and governance. Economic activity in much of the region is dominated by governments, while foreign fishing access agreements and foreign aid comprise significant components of national budgets.

While coastal fisheries provide important sources of traditional food and income to artisanal communities, the oceanic tuna fisheries are the cornerstone upon which many Pacific island states depend for revenue and economic activity. Fortunately, the region is home to the world's richest and largest tuna fishery with an estimated value of approximately AU$3.9 billion in 2007 (Williams and Terawasi, 2008).

These fisheries have long been viewed as the primary development opportunity for many of the region's developing island states. Access fees from foreign fishing vessels deliver much-needed financial contributions to governments, while domestically-based fishing fleets and support industries pump hard currency into national economies. Fisheries resources have also, to a degree, motivated some distant water fishing nations (DWFNs) to build and maintain relationships throughout the region that include significant aid budgets.

The four key tuna species of interest (albacore, skipjack, yellowfin and bigeye) migrate across the EEZs and high seas pockets throughout the region. Unlike Atlantic, Indian and Eastern Pacific tuna fisheries, the majority of fishing effort in the WCPO occurs within the EEZs of the Pacific island states, Indonesia and the Philippines. Approximately 57 per cent of all WCPO catches for the four key tuna species are taken from the Pacific island EEZs.[3]

The two main components of the Pacific islands tuna fisheries comprise distant water fishing vessels and domestic fishing vessels. Distant water fishing vessels may either be based within a Pacific island state (due to licensing requirements or operational factors) or operate from a distant home port. The vast majority of them are from distant water fishing nations (DWFN), notably China, Japan (Tarte, 1997), Korea, the United States, Taiwan and increasingly, the European Union, who fish within Pacific island EEZs or on the high seas. These vessels operate through access agreements or are directly licensed by the coastal states to fish within their EEZ. The annual value of tuna caught by DWFN vessels is approximately four times that caught by domestic fishing vessels (ForSEC, 2005a, b).

Domestic fishing vessels are generally smaller ones that fish for tuna within their own flag state's EEZ. They may be nationally owned and operated, or may be foreign owned and operated through domestic charters and/or joint ventures with local interests. Charter and/or joint venture arrangements generally specify local participation requirements in the venture and require that the vessel be located within the country. Most domestic vessels are longliners, but recently there has been an increase in Pacific island flagged or domestic-based purse seiners (ForSEC, 2005a, b).

In recent years, the problems of overfishing and overcapacity have increased and now threaten the long term sustainability of some of the region's key fish stocks. The WCPFC Scientific Committee has repeatedly expressed concerns about current levels of fishing since its inaugural meeting in 2005 and each year recommends tougher reductions (WCPFC Secretariat, 2005, 2006, 2007 and 2008). Furthermore, economic studies have shown that fishing effort is significantly above optimal levels, thereby reducing the profitability of the fishery and undermining opportunities for Pacific island states to develop fishing and related industries (Bertignac *et al.*, 2001; Parris and Grafton, 2005; Reid *et al.*, 2003; Kompass and Tuong, 2006).

Table 1: Access fees and aid

| | Aid<br>% of GDP | Fish access fees<br>% of GDP | Government expenditure<br>% of GDP | Government employment<br>% of total paid employees |
|---|---|---|---|---|
| Fiji | 2.3 | 0.01 | 19.4 | 18.7 |
| Papua New Guinea | 6.4 | 0.17 | n/a | n/a |
| Solomon Islands | 25.5 | 0.1 | 53.1 | 32.5 |
| Vanuatu | 11.7 | 0.1 | 39.4 | 33.3 |
| Cook Islands | 3.5 | 0.21 | n/a | n/a |
| Samoa | 10.4 | 0.08 | 28 | 10 |
| Tonga | 16.3 | 0.1 | 43.4 | 41 |
| Tuvalu | 38.6 | 42.6 | n/a | n/a |
| Kiribati | 31.5 | 42.81 | 100.6 | 28.7 |
| Marshall Islands | 53.9 | 5.12 | 98.1 | 46.4 |
| Fed. St. Micronesia | 49.5 | 6.7 | 89.8 | 69.2 |
| Nauru | 35.5 | 6.59 | n/a | n/a |
| Palau | 20.6 | 0.7 | n/a | n/a |

Sources: World Bank, 2002; Gillet and Lightfoot, 2001; AusAID, 2006

The Pacific island states depend upon regional cooperation and regional institutions to enable and support effective fisheries management and development. The success of these institutions is critically important given the region's high dependence upon fisheries resources and their limited national capacity. Any serious threat to the sustainability of the tuna resource can be viewed as a threat to the region's economic viability and food security.

## Regional Approaches to Fisheries Cooperation and Capacity Building

The Pacific Islands Forum Fisheries Agency (FFA) has played the central role in fostering regional cooperation amongst its membership in their management and development of the region's tuna fisheries. Based in Honiara, Solomon Islands, the FFA was founded in 1979 by the independent Pacific island states, Australia and New Zealand and sits within the Pacific Islands Forum umbrella. In 1979, the independent members of the Pacific Islands Forum (then named the South Pacific Forum) foresaw the challenges involved in managing and developing their newly proclaimed EEZs and recognised that individually they did not have the capacity to respond adequately. With remarkable vision they combined their resources and established the FFA to promote intra-regional cooperation and harmonisation of fisheries management policies. The mission of the FFA is to support and enable Pacific island states to achieve sustainable fisheries and maximise their social and economic benefits in harmony with the broader environment (FFA, 2005).

The FFA itself does not manage the tuna fisheries and has no such mandate, nor any authority to enforce decisions of its governing council (Aqorau, 2002). This is an important distinction as the establishment of the FFA caused some concern amongst DWFN that the FFA was a regional fisheries management organisation and therefore should be open to the participation of distant water fishing states with an interest in the tuna fisheries (Van Dyke and Heftel, 1981; Van Dyke, 1995). These concerns reflected the interests of some DWFNs, particularly the USA, that did not necessarily recognise the sovereign rights of coastal states over migratory fisheries (Aqorau, 2002), or properly recognise the fundamental capacity building purpose of the FFA.

The FFA supports the interests of the Pacific island states through facilitating regional cooperation in their favour and providing technical and policy advice. Concerns over the role of the FFA were effectively laid to rest in the early 1990s[4] and discussions began to develop momentum for the establishment of the WCPFC.

## Cooperative Approaches to Fisheries Science and Data Analysis

The FFA works closely with its partner agency, the Secretariat of the Pacific Community (SPC), to facilitate regional cooperation and support the management and development of the region's tuna fisheries, at various national, sub-regional and regional levels.

SPC, formerly the South Pacific Community, was the first of the regional fora to be established and was founded in 1947 by the colonial powers of the time: Australia, New Zealand, Netherlands, France, United Kingdom (UK) and the United States of America (USA). The membership evolved through the period of decolonisation and now includes the independent Pacific island states,[5] the Pacific island territories[6] and Australia, New Zealand, France and the USA. The organisation is headquartered in Noumea, with regional offices throughout the Pacific islands region. Its mission is to '... help Pacific island people make and implement informed decisions about their future' (SPC, 2005).

The Oceanic Fisheries Programme (OFP) is one of a number of SPC programmes that aim to build capacity within the Pacific islands region and support members with technical assistance. The mission of the OFP is: 'to provide member countries with the scientific information and advice necessary to rationally manage fisheries exploiting the region's resources of tuna, billfish and related species' (SPC, 2005). The OFP provides fisheries science services to its members (primarily relating to tuna) and is also a contracted science provider for the Scientific Committee of the Western and Central Pacific Fisheries Commission.

## Cooperative Approaches to Fisheries Management and Domestication

The FFA has been most successful in its work to support sub-regional and regional cooperation relating to access by foreign fishing fleets into EEZs. In this area, the FFA has facilitated the development of a number of key regional arrangements. The framework for subsequent successes in Pacific island fisheries cooperation was established in 1982 by a subset of the FFA membership who have since become the driving force within the FFA, and consequently have benefited most from regional cooperation. The 1982 Nauru Agreement

Concerning Cooperation in the Management of Fisheries of Common Interest (Nauru Agreement, 1982) was negotiated by the equatorial Pacific island states whose waters include the most significant fisheries.[7]

The Pacific island parties to the Nauru Agreement (PNA) recognised that they were in a weak position when negotiating access arrangements individually with DWFNs, particularly when DWFNs played each state against each other in negotiations over access fees and conditions (Lodge, 1992). In response, the PNA negotiated the Nauru Agreement in order to coordinate and harmonise their fisheries management and access conditions, thereby placing themselves in a stronger strategic position when negotiating with DWFNs. The Nauru Agreement promoted the following objectives:

- Coordinate and harmonise management of common fish stocks between PNA, without derogating any of their sovereign rights (Article 1);
- Priority consideration for licensing PNA vessels over foreign vessels (Article 2a);
- Establish minimum terms and conditions for foreign vessel access (Article 2b);
- Cooperate and coordinate fisheries monitoring, control and surveillance (Articles 6 and 7).

The Nauru Agreement became the cornerstone for regional cooperation and enabled subsequent cooperative agreements to develop increasingly harmonised approaches to common fisheries that would extend beyond the limited membership of the PNA.

During the late 1980s and early 1990s, the Pacific island members of the PNA became increasingly concerned at the rapid expansion of the purse seine fishery and its potential impact on the long term sustainability of the WCPO tuna fisheries. In light of these concerns, PNA initiated discussions in 1990 to develop arrangements that might limit purse seine numbers within the PNA sub-region. During these discussions, PNA agreed to introduce interim limits on how many purse seine vessels they would license to fish in their collective EEZs while negotiating a more comprehensive arrangement to limit purse seine fishing across all PNA EEZs. In 1990, PNA agreed to provisionally limit the number of purse seine vessel licenses to 164 purse seine vessels but by 1993, this limit had increased to its final maximum of 205 (Dunn *et al.*, 2006). These increases were due to pressure from DWFN to license vessels and problems verifying exactly how many purse seine vessels were actually licensed across all PNA EEZs.[8]

In 1993, the Pacific island members of the PNA signed the legally binding Palau Arrangement for the Management of the Purse Seine Fishery in the Western and Central Pacific (Palau Arrangement, 1995) which subsequently entered into force in 1995. Prior to the establishment of the WCPFC, the Palau Arrangement was the only mechanism available to control purse seine fishing effort in the WCPO.

The Palau Arrangement aims to protect tuna stocks from overfishing and improve the economic benefits to Pacific island members of the PNA from access fees and fisheries development. It primarily does this through limiting the licenses available to fish within the PNA EEZs (therefore limiting catches and hopefully increasing prices) and enabling further cooperation in management of the purse seine fisheries between PNA. Given its exclusive coastal state membership, the scope of the Arrangement was effectively limited to EEZs. Significantly, the preamble to the arrangement did emphasise the special interest of coastal states in tuna in adjacent high seas areas.

## Cooperative Approaches to Fisheries Monitoring, Control and Surveillance

Immediately after concluding the Nauru Agreement, the Pacific island members of the PNA began negotiating the first of three implementing arrangements that would operationalise the treaty's objectives.[9] The First Implementing Arrangement to the Nauru Agreement (1IA, 1983) was adopted in September 1983 and established agreed Harmonised Minimum Terms and Conditions for foreign fishing vessels (HMTCs). While these conditions were originally intended to only apply to PNA, the broader FFA endorsed a draft of the conditions during their negotiations and began a parallel initiative that quickly extended the application of the HMTCs to the entire FFA membership. The HMTCs harmonised licensing procedures and catch reporting and established a regional register of fishing vessels. Each Pacific island is responsible for the implementation of these conditions at the national level (Aqorau, 2002).

In April 1990, following a significant increase in the number of vessels fishing in PNA waters and a desire from some PNA to review the HMTCS, the PNA commissioned a legal drafting group to prepare a draft second implementing arrangement. The Second Implementing Arrangement to the Nauru Agreement (2IA, 1991) came into effect in January 1991 and expanded the HMTCs to also incorporate observer requirements, prohibit transhipments at sea, expand monitoring and surveillance, and introduce an annual registration for the regional vessel register. Once again the broader membership of the FFA endorsed the PNA's expanded HMTCs and agreed that the conditions should be implemented throughout all FFA EEZs (Forum Communiqué, 1990).

Regardless of these new agreements on harmonised licensing conditions, Pacific island states still suffered from an obvious lack of capacity to patrol and monitor their massive EEZs. In response, the Pacific island members of the FFA adopted a treaty framework in 1993[10] that enabled FFA member states to cooperate in surveillance and enforcement and share surveillance assets.

In a similar light, the FFA further built on these precedents by agreeing to establish the world's first centralised satellite based vessel monitoring system. In 1997, the entire FFA membership of Pacific island states agree to expand the HMTCs and require all their licensed foreign vessels to report continuously to a satellite based vessel monitoring system (VMS) that would be operated by the FFA secretariat and would forward vessel positions to national officers to monitor.

A decade later, in May 2008, the Pacific island members of the PNA updated their requirements for licensed foreign fishing vessels and introduced new additional licensing terms and conditions that set global precedents in coastal state management of tuna fisheries. Firstly, the PNA agreed that they would collectively apply additional licensing terms and conditions that introduced new conservation and management requirements within their EEZs. However, more significantly, the PNA agreed that they would also prohibit licensed fishing vessels from fishing in two high seas pockets surrounded by PNA EEZs north and northeast of Papua New Guinea.

Fishing vessels may voluntarily accept or refuse this new licensing condition. Vessels may continue fishing on the high seas if they wish, but in so doing they may not fish in PNA EEZs. As such, this does not breach the freedom of the seas that is enshrined in Article 87 of the UN Convention on the Law of the Sea (UNCLOS, 1994). However, given that the PNA EEZs contain the most productive fishing grounds, it is a powerful tool and quickly raised the concern of DWFN. Despite significant opposition from DWFN interests, the PNA signed the Third Implementing Arrangement (3IA, 2008) in Palau in May 2008.

## Regional Approaches to Cooperation between Pacific Island States and Distant Water Fishing Nations (DWFN)

The limited coastal state membership of the FFA and PNA inevitably limit their effectiveness. Until recently, fishing effort targeting the same migratory stocks on the high seas and inside the neighbouring waters of Indonesia and the Philippines was essentially unregulated. In the early 1990s, FFA members recognised that a regional forum was required to engage their DWFN partners and Indonesia and Philippines and enable management of migratory fisheries beyond their EEZs.

In 1994, the FFA hosted a multilateral high level conference of Pacific island states and DWFN on the future management and conservation of straddling and highly migratory fisheries within the WCPO. This meeting agreed on the need to cooperatively and sustainably manage WCPO tuna resources across their entire range.

This was followed by six further conferences until negotiations concluded in 2000 with the successful adoption of the Western and Central Pacific Fisheries Convention[11] (WCPF Convention, 2000) which subsequently entered into force in July 2004. The objective of the WCPF Convention, as described in Article 2, is to ensure the long term conservation and sustainable use of WCPO straddling and highly migratory fish stocks in accordance with the 1982 Convention (LOSC) and the Agreement (UNFSA). The Convention establishes the decision making Western and Central Pacific Fisheries Commission (WCPFC), which meets annually, and a secretariat which is headquartered in the Federated States of Micronesia.

The Pacific island states are a critical membership bloc of the WCPFC and were a key driver behind its development. Other WCPFC members include (amongst others) Indonesia, Philippines and the DWFNs: Japan, Korea, China, Taiwan, USA and the European Community. The WCPF Convention binds these members to implement its provisions and WCPFC conservation and management measures. Since its establishment in 2004, the WCPFC has agreed on a number of conservation measures that impose specific obligations on all members.

The WCPFC closely follows the framework established by the UN Fish Stocks Agreement (UNFSA, 1995) and emphasises a precautionary and ecosystem based approach to fisheries management. The WCPF Convention applies to all waters of the WCPO, including both high seas and EEZs. However, the WCPF Convention clearly states in Article 4 that nothing in the Convention shall prejudice the rights, jurisdiction and duties of states under the LOSC and UNFSA, and that the WCPFC shall be interpreted and applied in the context of, and in a manner consistent with the UNCLOS and UNFSA. This is a critical point for Pacific island states given their heavy dependence upon the fishery and aspirations for development, and their sovereign rights over much of the fishery within their EEZs.

A critical focus for the WCPFC will be how it develops co-operative management across the high seas/EEZ nexus, and by operation or intent, allocates rights to the tuna resource. There are key disagreements between DWFN and Pacific island states over how the WCPF Convention should be interpreted regarding implementation of management measures in EEZs and on the issue of allocation. Both interpretations cite relevant articles of the WCPF Convention and UNFSA (Aqorau, 2001).

Pacific island states argue that the main purpose for the WCPFC is to regulate the high seas and ensure that stocks are not over-fished there (Ram-Bidesi and Tsamenyi, 2004). They note that management measures already exist within their EEZs. This argument is supported by provisions within both the WCPFC and the UNFSA which require measures to be compatible across the high seas/EEZ nexus, taking into account existing measures already in practice.

DWFN argue that the WCPFC, as the primary management authority for tuna across the WCPO, should establish management and conservation measures across the entire range of the stocks, both inside EEZs and on the high seas (Cordonnery, 2000). These states refer to Article 10 of the WCPFC which provides that the Commission can determine the quantity of catches, levels of effort, limitations on fishing capacity and other necessary management measures throughout the convention area.

Pacific island states respond that the Commission can establish 'global' catch, effort and/or capacity limits across the entire Convention area, but that it is the sovereign right of coastal states to determine catches within their EEZs. This is supported by the 'without prejudice to the sovereign rights of coastal states' clause in Article 10 of the WCPF Convention regarding the Commission's functions.

Resolving these conflicts will be critical to the effective functioning of the WCPFC and its ability to agree upon, and implement effective conservation and management measures across the range of the stocks.

## Lessons and Insights for the Commonwealth

The past 30 years have demonstrated a remarkable level of cooperation within the Pacific islands region that has substantially increased the capacity of the region to manage their fisheries and successfully negotiate with far more powerful DWFNs, in particular the USA and Japan.

Analysis of these cooperative approaches reveals some lessons and insights for other developing states throughout the Commonwealth, particularly in the Caribbean, West Africa and Southern Africa. Three areas of cooperation are of particular interest: science and management; access; and influence.

Firstly, in the area of science and management, the Pacific islands region has established a number of cooperative precedents that could productively be explored and perhaps emulated in other fisheries dependent regions where developing states have little capacity and share an interest in straddling or migratory stocks. Furthermore, some useful lessons have been learnt that may help other developing states throughout the Commonwealth to better manage and develop their fisheries.

While much has been achieved by the Pacific islands region, until recently conservation has not been a key priority. Unfortunately, analysis by the OFP shows that overfishing of bigeye and possibly yellowfin is occurring and that a minimum 30 per cent reduction is required to halt overfishing. In early 2008, the OFP suggested that management responses had so far been ineffective at limiting fishing effort and capacity to sustainable levels and that further management responses were required, in particular by reducing the impact of fish aggregating devices (FADs) on juvenile populations and reducing over-fishing within the Philippines and Indonesia (SPC, 2008).

A key weakness of the Pacific island region's fishing vessel cap was that it did not account for effort creep (where fishing vessels catch ever more fish as technology improves, engines get more powerful, nets get bigger). While the fishing vessel cap of 205 remained stable, it was increasingly seen as a blunt and not particularly effective tool at promoting conservation and development interests. Problems emerged as the fishing vessel cap made it difficult for new fleets to enter the fishery. A more effective mechanism was required that would enable better implementation of conservation goals and support the short and long term development interests of PNA members (Rodwell, 2004).

In 2007, the Pacific islands members of the PNA reviewed the vessel cap and agreed to introduce a limit on the number of purse seine days. Vessel days could be sold in such a way as to maximise economic returns and would introduce greater fleet flexibility and better enable conservation outcomes.

The Vessel Day Scheme (VDS) was introduced in December 2007 and aimed to constrain catches to sustainable levels and increase benefits from fishing activities through access fees paid by DWFNs. The VDS replaced the broad purse seine vessel number cap with a set number of days that can be fished in the combined EEZs of the PNA. Vessel days are then allocated to each PNA. The VDS enabled PNA to account for effort creep by differentiating fishing days based on vessel length and allowing for vessel formulas to be modified over time to account for changes in technology and efficiency.

Secondly, on the matter of access by foreign fleets to the fisheries of developing states, the Pacific islands region has taken a proactive approach to developing its own fisheries and progressing its own aspirations. For example, a key objective of the VDS is to create competition between DWFN vessels to purchase fishing days at the maximum price. As the VDS has been introduced, allowances have been made for vessels that fish under an agreement between the USA and the Pacific island states (the USMLT)[12] and an internal arrangement within the PNA membership that supports regional development aspirations; the FSM Arrangement.[13]

However, these advances also created some problems. The current USMLT restricts the number of USA vessels that can be licensed under the USMLT to 45 vessels. However, the number of vessels licensed under the treaty declined to a low of 11 in 2007 due to economic factors. Since then, vessel numbers have dramatically increased for two reasons. First, the USA removed its restriction on vessel origin and now allows foreign built (and significantly cheaper) vessels to be licensed under the treaty. Second, the imminent implementation of fishing limits by the equatorial Pacific island states created an incentive for vessels to come under the USMLT umbrella to avoid restrictions on fishing effort (FFA, 2008).

On a separate note, two trends emerged through the FSM Arrangement. First, it appears to have spurred some PNA to invest in the harvesting sector. Second, it has encouraged foreign direct investment into PNA to obtain fishing licenses for all PNA waters (Havice, 2007). However, for various reasons, the FSM Arrangement seems to have had limited success in domesticating vessels from DWFN into becoming truly locally based (or domestic) in the PNA. A review of the FSM Arrangement in 2007 found that there was significant variation in the amount and quality of benefits gained from the FSM Arrangement to PNA and that other factors, not related to the FSM Arrangement, were probably just as important as the Arrangement in promoting domestication and onshore development (i.e proximity to fishing grounds, availability of land, infrastructure and services and domestic government policy). (Parris *et al.*, 2007).

Lastly, on the matter of influence, the Pacific islands region has significantly boosted its collective and individual capacity to stand up to DWFN delegations and pressure and have recently expanded their influence into the immediately adjacent high seas. For example, in 2006 and 2008, key arguments between coastal states and DWFN were partly resolved in practice (though not clearly in principle) through the incorporation of the PNA VDS and the PNA 3IA into WCPFC conservation and management measures. These decisions indirectly recognised the primacy of coastal states over management of fisheries within their EEZs and framed conservation and management for high seas fisheries in the context of existing management practised in EEZs. A key example of this was the endorsement of the PNA 3IA's closure of the high seas pockets and its inclusion within the WCPFC bigeye and yellowfin conservation and management measure (CCM, 2008–01). It is highly unlikely that the WCPFC would have agreed to close any high seas areas without their hand being previously forced by the PNA 3IA.

While there was progress on the issue of EEZ/High Seas compatibility, coastal state/DWFN tensions arose in regard to the application of the WCPFC to the archipelagic waters of coastal states. Due to some controversy over the issue, the Chair of the WCPFC requested an opinion from the legal counsel on, amongst other things, the application or otherwise of the WCPFC to archipelagic seas. The legal counsel referred to the WCPF Convention, UNCLOS and UNFSA and suggested that the WCPF Convention only has application to the high seas and EEZs, and not the internal waters, archipelagic waters and territorial seas, due to qualifications in UNFSA and the Convention between 'sovereign rights' and 'sovereignty'. Nevertheless, the legal counsel noted that in addition to the WCPF Convention, the UNCLOS and UNFSA, other principles of international law need to be considered, particularly the principle of 'good neighbourliness' which requires that states must act in good faith and ensure that activities in their territories do not cause harm or affect the interests of other states.

Despite this, the USA continued to argue against such an interpretation and stated its position that the WCPFC applies to archipelagic waters, as well as EEZs and high seas. Confusingly, the USA distinguished between territorial seas and archipelagic seas and stated its opinion that territorial seas remained excluded from the application of the WCPF Convention (WCPFC Summary Report, 2008). Pacific island states, Philippines and Indonesia refuted the USA position and argued that archipelagic waters are deemed to be sovereign territory of the coastal state and as such as outside the jurisdiction of the WCPFC.

## Conclusion

Recent achievements within the WCPFC, particularly the bigeye and yellowfin conservation measure, demonstrate the strength of the FFA and PNA sub-group when they negotiate collectively. This sends an important signal to other developing states and regions throughout the Commonwealth that there is much to be gained from such cooperative approaches.

Similarly, the achievements of the FFA and PNA management, control and development mechanisms demonstrate the potential of this sub-regional grouping to manage fishing efforts throughout its area in the direct interests of its members, and to extend its influence beyond its immediate boundaries. While neither the current VDS nor WCPFC conservation and management measures yet meet conservation requirements as recommended by SPC and the WCPFC Scientific Committee, they provide the initial framework – due almost entirely to the drive of the FFA and PNA.

Given the high dependence of the region on fisheries resources for revenue and food security, it is vital that these forums achieve their conservation, management and development goals and enable the Pacific island states to implement the institutional and governance programs necessary to conserve and develop the WCPO tuna fisheries.

As with many developing state regions throughout the Commonwealth, particularly within the Caribbean and Africa, fisheries is a key resource and one that is difficult to manage. As pressures grow, and global fishing fleets become more aggressive in their hunt for open fishing grounds, strong cooperative institutions will become increasingly critical to the effective management, development and control of regional fisheries.

## References

1IA. (1983) Full title: An Arrangement Implementing the Nauru Agreement Setting Forth Minimum Terms and Conditions of Access to the Fisheries Zones of the Parties. Copy available in Appendix 2 of: Lodge. (1992). Minimum Terms and Conditions of Access: Responsible Fisheries Management Measures in the South Pacific Region. *Marine Policy*, July 1992, pp. 277–305.

2IA. (1991) Full title: A Second Arrangement Implementing the Nauru Agreement Setting Forth Additional Terms and Conditions of Access to the Fisheries Zones of the Parties. Copy available in Appendix 2 of: Lodge. (1992). Minimum Terms and Conditions of Access: Responsible Fisheries Management Measures in the South Pacific Region. *Marine Policy*, July 1992. pp. 277–305.

3IA. (2008) Full title: A Third Arrangement Implementing the Nauru Agreement Setting Forth Additional Terms and Conditions of Access to the Fisheries Zones of the Parties. Forum Fisheries Agency (FFA). Honiara. Solomon Islands.

Aqorau, Transform (2000) Illegal Fishing and Fisheries Law Enforcement in Small Island Developing States: the Pacific Islands Experience. *International Journal of Marine and Coastal Law*, 15 (1), pp. 37–61.

Aqorau, Transform. (2001) Tuna fisheries management in the Western and Central Pacific Ocean: A critical analysis of the Convention for the Conservation and Management of Highly Migratory Fish Stocks in the Western and Central Pacific Ocean and its implications for the Pacific SIDS States, *International Journal of Marine and Coastal Law*, 16 (3), pp. 379–431.

Aqorau, Transform. (2002) *Cooperative Management of Shared Fish Stocks in the South Pacific*. Paper Presented at the Norway-FAO Expert Consultation on the Management of Shared Fish Stocks – Bergen, Norway, 7–10 October 2002.

AusAID. (2006) Pacific 2020. *Challenges and Opportunities for Growth*. Australian Agency for International Development (Canberra: AusAID).

Bertignac, M., Campbell, H., Hampton, J. and Hand, A. (2001) Maximising Resource Rent from the Western and Central Pacific Tuna Fisheries. *Marine Resource Economics*, 15. pp. 151–177.

Cartwright, I. and Willock, A. (1999) *Oceana's Birthright; the role of rights-based management in tuna fisheries of the Western and Central Pacific*. Paper presented to the FishRights 99 Conference, Perth. Australia. 11–19 November 1999.

Cordonnery, L. (2002) A note on the 2000 Convention for the Conservation and Management of Tuna in the Western and Central Pacific Ocean, *Ocean Development and International Law*, 33. pp. 1–15.

Doulman, D. (1987) Fisheries Cooperation: The Case of the Nauru Group. In *Tuna Issues and Perspectives in the Pacific Islands Region* edited by Doulman, David (Honolulu:East West Centre).

Dunn, S., Rodwell, L. and Joseph, G. (2006) *The Palau Arrangement for the Management of the Western Pacific Purse Seine Fishery – Management Scheme* (Vessel Day Scheme). Paper Presented at the Sharing the Fish Conference, March 2006, Perth Western Australia. *http://www.fish.wa.gov.au/docs/events/ShareFish/papers/pdf/papers/GlenJoseph.pdf*

FFA (2005) FFA Strategic Plan 2005–2020 (Honiara).

FFA (2008) *Internal Briefing Paper for Parties to the Nauru Agreement on issues relating to the USMLT and Vessel Day Scheme*. February 2008 (Honiara).

ForSEC (2005a) *Enabling Environment – Good Governance and Security*. Pacific Plan Regional Analysis Papers (Suva. Pacific Islands Forum Secretariat).

ForSEC (2005b) *Fisheries*. Pacific Plan Regional Analysis Papers. Pacific Islands Forum Secretariat. Accessed 18 December 2007. *http://www.pacificplan.org/tiki-list_file_gallery.php?galleryId=11*

Forum Communiqué (1990) Forum Communique of the Twenty-First South Pacific Forum. Port Vila. Vanuatu. 31 July–1 August 1990. Accessed online 19 March 2009. *http://www.forumsec.org/_resources/article/files/1990 per cent20Communique.pdf*

FSM Arrangement (1995) FSM Arrangement for Regional Fisheries Access. Accessed online 11 March 2009. *http://www.ffa.int/node/30#attachments*

Gillett, R. (2005) Pacific Islands Region. *Review of World Marine Fishery Resources*, 457. pp. 144–157 (Rome. United Nations Food and Agriculture Organisation).

Gillet, R. and Lightfoot, C. (2001) *The Contribution of Fisheries to the Economies of Pacific Island Countries*. Report prepared for Asian Development Bank, Forum Fisheries Agency and World Bank.

Havice, E. (2007) *The State of Play of Access Agreements with Distant Water Fishing Partners: Implications and Options for Pacific Island Countries* (Honiara: Forum Fisheries Agency).

Kompass, T. and Tuong, N.C. (2006) Economic Profit and Optimal Effort in the Western and Central Pacific Tuna Fisheries, *Pacific Economic Bulletin* 21(3), pp. 46–62.

Lodge (1992) Minimum Terms and Conditions of Access: Responsible Fisheries Management Measures in the South Pacific Region. *Marine Policy*, July 1992, pp. 277–305.

Lodge (1998) The Development of the Palau Arrangement for the Management of the Western Pacific Purse Seine Fishery. Marine Policy, 22 (1), pp. 1–28.

Molenaar, E.J. (2003) Participation, Allocation and Unregulated Fishing: The Practice of Regional Fisheries Management Organisations. *The International Journal of Marine and Coastal Law*, 18 (4) pp. 457–820.

Nauru Agreement (1982). Full title: The 1982 Nauru Agreement Concerning Cooperation in the Management of Fisheries of Common Interest. Accessed online 10 March 2009 at *http://www.ffa.int/node/93#attachments*

Niue Treaty (1993) The Niue Treaty on Co-operation in Fisheries Surveillance and Law Enforcement in the South Pacific Region. Reprinted in *Commonwealth Law Bulletin*. 702, 1993, p. 32. International Legal Materials.

Palau Arrangement (1995). The Palau Arrangement for the Management of the Purse Seine Fishery in the Western and Central Pacific. Accessed online 10 March 2009 at *http://www.ffa.int/node/91#attachments*

Parris, H. and Grafton, Q. (2005) Can Tuna Promote Sustainable Development in the Pacific. *Journal of Environment and Development* 15, pp. 259–96.

Parris, H., Barclay, K. and Cartwright, I. (2007) *Review of the Federated States of Micronesia Arrangement*. Forum Fisheries Agency (Honiara).

Petersen, E. (2003) The Catch in Trading Fishing Access for Foreign Aid, *Marine Policy*, 27. pp. 219–228.

Ram-Bidesi, V. and Tsamenyi, M. (2004). Implications of the tuna management regime for domestic industry development in the Pacific SIDS States. *Marine Policy*, 28, pp. 383–392.

Reid, C., Squires, D., Joen, Y., Rodwell, L. and Clark, R. (2003). Analysis of Fishing Capacity in the Western and Central Pacific Ocean Tuna Fishery and Management Implications. *Marine Policy* 27, pp. 229–469.

Rodwell, L. (2004) *FFA Initiatives related to the Palau Arrangement, Purse Seine Management and the Management of Bigeye Fishing Mortality in the WCPO*. Paper presented to the 17th Meeting of the Standing Committee on Tuna and Billfish. Majuro, Marshall Islands. 8–18 August. 2004. SCTB17 Working Paper. FTWG-6.

SPC (2005). Secretariat of the Pacific Community Vision and Mission. *http://www.spc.int/corp/index.php?option=com_content&task=view&id=22&Itemid=73*

SPC (2008). Update on Tuna Fisheries. Secretariat of the Pacific Community. 1/2008. *http://www.spc.int/oceanfish/Docs/Tuna per cent20Fisheries_1-2008.pdf*

Tarte, S. (1997) *Diplomatic Strategies: The Pacific Islands and Japan*. Pacific Islands Economic Paper, 269. July 1997 (Australia-Japan Research Centre).

UNCLOS (1994) United Nations Convention on the Law of the Sea, signed at Montego Bay on 10 December 1982, *International Legal Materials*, 21 (1982), pp. 1261–1354.

UNDP and FFA (2004) *Pacific Islands Oceanic Fisheries Management Project*. 2004. (Honiara: United Nations Development Programme and Forum Fisheries Agency).

UNFSA (1995) Full title is Agreement for the Implementation of the Provisions of the United Nations Convention on the Law of the Sea of 10 December 1982 Relating to the Conservation and Management of Straddling Stocks and Highly Migratory Fish Stocks. Signed December 1995. New York, USA. Entered into force 2001. Accessed online April 2005 at *http://www.un.org/Depts/los/convention_agreements/convention_overview_fish_stocks.htm*

USMLT (1988). Full title: Treaty on Fisheries Between the Governments of Certain Pacific Island States and the Government of the United States of America. Opened for signature April 2. 1987. Reprinted in 26 *International Legal Materials* 1048. 1987.

Van Dyke, J. (1995). Modifying the 1982 Law of the Sea Convention: New Initiatives on Governance of High Seas Fisheries Resources: the Straddling Stocks Negotiations. *The International Journal of Marine and Coastal Law*, 10 (3), pp. 219–227.

Van Dyke, J. and Heftel, S. (1981) Tuna Management in the Pacific: An Analysis of the South Pacific Forum Fisheries Agency. *University of Hawaii Law Review* 3, pp. 1–64.

WCPF Convention (2000) Full title: is *Convention for the Conservation and Management of Highly Migratory Fish Stocks in the Western and Central Pacific Ocean*. Signed September 2000. Honolulu, USA. Entered into force, 2004. Accessed online 10 July 2006 at *http://www.wcpfc.int/*

WCPFC Secretariat (2005) The Commission for the Conservation and Management of Highly Migratory Fish Stocks in the Western and Central Pacific Ocean. Scientific Committee. Fourth Regular Session, 8–19 August 2008, Noumea, New Caledonia.

WCPFC Secretariat (2006) The Commission for the Conservation and Management of Highly Migratory Fish Stocks in the Western and Central Pacific Ocean. Scientific Committee. Fourth Regular Session, 7–18 August 2008, Manila, Philippines.

WCPFC Secretariat (2007). The Commission for the Conservation and Management of Highly Migratory Fish Stocks in the Western and Central Pacific Ocean. Scientific Committee. Fourth Regular Session, 13–24 August 2008, Honolulu.

WCPFC Secretariat (2008). The Commission for the Conservation and Management of Highly Migratory Fish Stocks in the Western and Central Pacific Ocean. Scientific Committee. Fourth Regular Session, 11–22 August 2008, Port Moresby, Papua New Guinea.

WCPFC Summary Report (2008) Summary Report of the Commission for the Conservation and Management of Highly Migratory Fish Stocks in the Western and Central Pacific Ocean. Fifth Regular Session, 8–12 December 2008. Pusan, Korea.

WCPFC CMM 2008-04 Conservation and Management Measure to prohibit the use of large scale driftnets on the high seas in the convention area. Fifth Regular Session of the Western and Central Pacific Fisheries Commission. Busan, Republic of Korea. 8–12 December 2008.

Williams, P. and Terawasi, P. (2008) *Overview of Tuna Fisheries in the Western and Central Pacific*

*Ocean, including Economic Conditions – 2007*. Paper presented to the Fourth Regular Session of the Scientific Committee of the Western and Central Pacific Fisheries Commission. 11–22 August 2008. Port Moresby, Papua New Guinea. WCPFC-SC4-2008/GN WP-1.
World Bank (2002) Pacific Islands Regional Economic Report, World Bank, Washington, DC.

## Notes

1 For the purposes of this paper, the term Pacific island states refers to the FFA membership, excluding Australia and New Zealand. This includes: Cook Islands, Federated States of Micronesia, Fiji Islands, Kiribati, Marshall Islands, Nauru, Niue, Palau, Papua New Guinea, Samoa, Solomon Islands, Tokelau, Tonga, Tuvalu, Vanuatu.
2 For the purposes of this paper, the WCPO is defined as those waters within the area defined by the Convention on the Conservation and Management of Highly Migratory Fish Stocks in the Western and Central Pacific Ocean. This stretches from Indonesia and the Philippines in the west to Hawaii, Kiribati and French Polynesia in the East.
3 For the purposes of this estimate, this includes the EEZs of: (FFA members) Cook Islands, Federated States of Micronesia, Fiji, Kiribati, Marshall Islands, Nauru, Niue, Palau, Papua New Guinea, Samoa, Solomon Islands, Tokelau, Tonga, Tuvalu, Vanuatu (and non-FFA members) American Samoa, French Polynesia, New Caledonia, Pitcairn Island, and the French territory of Wallis and Futuna. The data for this estimate was sourced from Gillett, 2005.
4 While some DWFN (most notably the USA) historically rejected interpretations of the UN Convention on the Law of the Sea (UNCLOS) that granted coastal states sovereign rights over migratory species within their EEZs, the reality at sea moved on. For over 20 years, the *status quo* has reflected the coastal state interpretation and there is no indication that fishing states are likely to fish for tuna within EEZs without the permission of coastal states. It is now widely accepted that the UNCLOS grants coastal states 'practically exclusive powers over regulating access' to the fisheries within their EEZ, including straddling and highly migratory fish stocks (Molenaar, 2003).
5 Cook Islands, Federated States of Micronesia (FSM), Fiji, Kiribati, Marshall Islands, Nauru, Niue, Palau, Papua New Guinea (PNG), Samoa, Solomon Islands, Tonga, Tuvalu, Vanuatu.
6 American Samoa, French Polynesia, Guam, New Caledonia, Northern Mariana Islands, Pitcairn Islands, Tokelau, Wallis and Futuna.
7 Papua New Guinea, Federated States of Micronesia, Kiribati, Marshall Islands, Nauru, Palau, Solomon Islands were all original signatories. Tuvalu subsequently became a party in 1991.
8 The development of the Palau Arrangement caused much concern amongst DWFN and raised significant opposition. For discussion see Aqorau and Bergin, 1997; Lodge, 1998; Dunn *et al.*, 2006.
9 For details of negotiations for the First and Second Implementing Arrangement, see Lodge, 1992, and Doulman, 1987.
10 Niue Treaty on Co-operation in Fisheries Surveillance and Law Enforcement in the South Pacific Region entered into force in 1993 (Niue Treaty, 1993).
11 Full title is Convention on the Conservation and Management of Highly Migratory Fish Stocks in the Western and Central Pacific Ocean.
12 In 1987, the FFA negotiated a multi-lateral fisheries treaty between its members and the USA that recognised coastal state rights over migratory fisheries and significantly increased benefits to Pacific island states. The Treaty on Fisheries Between the Governments of Certain Pacific Island States and the Government of the United States of America (USMLT, 1988) commenced in 1988 and has since been renewed three times.

13 FFA interest in developing their own fisheries grew throughout the 1980s and 1990s and many Pacific island states aspired to replace DWFN fleets with locally based domestic fleets. In response to these aspirations, PNA members established the FSM Arrangement for Regional Fisheries Access (FSM Arrangement, 1995). The Arrangement further elaborated the Nauru Agreement's objectives of supporting local development and promoting PNA vessels over DWFN vessels. In this regard, the FSM Arrangement provided for lower cost licenses and access to the waters of all PNA states for domestic and locally based vessels that met specific criteria.

# 8 *Kawanatanga*: Fisheries Governance in New Zealand

Tom McClurg and Michel Arbuckle

## Introduction

Like every Commonwealth country, New Zealand has a unique story about the evolution of national governance over fisheries. This fact and New Zealand's implementation of a 'rights based' regime for commercial fisheries in recent years might suggest that New Zealand experiences have limited relevance to other jurisdictions. However, such a presumption does not bear examination. While the responses made by New Zealand to fisheries governance challenges are shaped by its particular natural, cultural, historical, and economic circumstances, the general situations in which these responses were formed are immediately familiar to other countries with a colonial heritage including much of the Commonwealth.

This chapter examines the changing role of the state in the governance of New Zealand fisheries over time. Compared with the subject of land, it is a topic that has received relatively little historical study. Fisheries governance has been a topic of marginal concern to the government and the population of New Zealand for most of the nineteenth and twentieth centuries. Recently, it has had a higher profile as the traditional role of the government in fisheries management was challenged by Maori. Old assumptions have been questioned and new approaches sought that are able to reconcile fisheries utilisation, fisheries sustainability and relations between races under the Treaty of Waitangi. In short, the challenge for fisheries governance in New Zealand is to develop a sophisticated post-colonial framework for fisheries management. This is a work in progress.

The history of New Zealand is that of a highly developed Polynesian tribal society impacted by European colonisation that imposed an English legal framework, along with predominantly British technology and attitudes to the use of unfamiliar natural resources. Although a negotiated framework for relations between Maori and European settlers of New Zealand was provided by the 1840 Treaty of Waitangi, the pressures of settlement led to a series of land wars that extended from 1845 until 1872. The defeat of Maori and the overtaking of Maori numbers by those of European ethnicity, led to a predominantly British perspective within the legislature of the nineteenth and twentieth centuries.

More recently, the national governance of New Zealand's fisheries has been carried out under the UN Convention on the Law of the Sea (UNCLOS). Consistent with international law, New Zealand established a 200 mile exclusive economic zone (EEZ) in 1977. Like many other Commonwealth countries, New Zealand is also a signatory of the FAO Code of Conduct for Responsible Fisheries (1995). Colonisation, its legacies and the contemporary framework of international law are fisheries governance factors that are certainly not unique to New Zealand.

In this context, governance means control or authority over fisheries exercised at a national level, or by the federal government of a country. Governance must therefore be distinguished from the activity of fisheries management which includes the handling and using of resources on a day to day basis. National governance concerns the role of the state in the design, implementation and enforcement of the institutions (rules) for fisheries use. It is common for states to extend their role from fisheries governance into fisheries management through the funding and delivery of research, administration and detailed regulatory activities. However, the problems of delivering good governance and good management are quite distinct and clear thinking about those roles is confounded if the differences between them are lost.

It is possible to see fisheries governance, fisheries management and fisheries use as a pyramid with three layers. Fisheries use (the base of the pyramid) is largely a private activity and fisheries governance (the apex) is a responsibility of government. This leaves fisheries management at the interface between the spheres of government and private interests. The appropriate location and configuration of this interface has been disputed at all times in New Zealand. While the government and officials have usually assumed that fisheries management is simply a government responsibility, Maori and ITQ owners assert that those who own rights to fish should have some responsibility for managing them.

## Maori Fisheries Governance Prior to 1840

Fisheries were crucial to the pre-European diet and economy of Maori. While in warmer parts of New Zealand, horticulture, especially the cultivation of sweet potato (*kumara*) was of great cultural significance, fish, shellfish and crustaceans were the most important source of protein. Although most fishing occurred close to shore, Maori also fished grounds well beyond the sight of land and had a vast knowledge of fish species, seasons of spawning and maturity, migratory patterns and fishing techniques including the use of hooks, nets and traps.

Maori society was organised into kin groupings with three main tiers. The largest group was the tribe (*iwi*) and major activities such as warfare were undertaken by *iwi* although there is also evidence of major fishing expeditions being organised and undertaken at this level. The iwi incorporated or exercised oversight over the rights of lesser kin groups within it. These are the extended family (*whanau*) and collections of *whanau* that form a sub tribe (*hapu*). *Whanau* held rights to particular small fisheries such as eel weirs, shell fish beds or fishing spots adjacent to their villages (*kainga*). Small canoes (*waka*) were also owned by *whanau*. *Hapu* exercised control over larger vessels, larger eel weirs and fishing grounds. Control of land was exercised predominantly at *hapu* level and this control extended to lakes, swamps and foreshore associated with that land.

Fisheries resources were therefore controlled by a system of exclusive rights exercised at a collective level within well defined geographical boundaries. These geographical boundaries extended into the sea. The management and use of resources within these areas was governed by chiefs (*rangatira*). The controls employed by *rangatira* constituted a body of law including seasonal closures and other prohibitions. The interpretation, promulgation and enforcement of these laws comprise part of *tino rangatiratanga*.

The governance situation prior to 1840 was described succinctly by Justice Greig in a High Court decision:

> *I am satisfied that there is a strong case that before 1840 Maori had a highly developed and controlled fishery over the whole coast of New Zealand at least where they were living. That was divided into zones under the control and authority of hapu and the tribes of the district. Each of these hapu and tribes had the dominion, perhaps the rangatiratanga, over those fisheries. Those fisheries had a commercial element and were not purely recreational or ceremonial or merely for the sustenance of local dwellers.*[1]

The temporal power of a *rangatira* was underpinned by their *mana*, a very broad and complex concept capturing inherited status, earned prestige and spiritual power. *Rangatiratanga* is an abstraction from the attributes of a chief: the many powers, qualities, responsibilities, whether exercised or implicit, of a chief. However, *rangatiratanga* is not a term applied at the level of an individual but the collective. It describes the process by which *iwi* and *hapu* determine what is right for them, that is to determine what will enhance their *mana*. Maori fisheries governance and Maori fisheries rights were therefore bound tightly together within cultural structures that encompassed political, economic, social and religious concepts and activities.

A powerful and ubiquitous concept of reciprocity (*utu*) supported controls over resource use and relations between individuals and groups. *Utu* required that favour be returned with favour and ill for ill. Failure to deliver appropriate *utu* undermined *mana*. The addition of European weapons to this cultural prerogative led to a series of intertribal wars, invasions and displacements known as the musket wars that continued from 1806 to 1845. The loss of life by Maori in these wars was historically unprecedented and considerably higher than the generally later wars between Maori and European (*Pakeha*). These violent convulsions meant that the pattern of Maori dominion over resources encountered by most settlers had frequently been established by recent conquests and was still subject to dispute.

## The Treaty of Waitangi 1840

New Zealand traces its history as a nation state from the signing of the Treaty of Waitangi. After the initial signing between representatives of Queen Victoria and assembled Maori chiefs at Waitangi in the Bay of Islands on 6 February 1840, copies of the Treaty were taken around New Zealand until over 500 Maori signatures were appended. The Treaty has three brief articles in both English and Maori but translation differences between the versions have provided a rich source of debate.

In Article One, chiefs ceded sovereignty over their respective territories to her Majesty the Queen of England. 'Sovereignty' was translated as *kawanatanga* (a transliteration meaning to have the attributes of a governor).

In Article Two, the Queen confirmed and guaranteed

> *... to the Chiefs and Tribes of New Zealand and the respective families and individuals thereof the full exclusive and undisturbed possession of their Lands and estates Forests Fisheries and other properties which they may collectively or individually possess so long as it is their wish and desire to retain the same in their possession.*[2]

In Article Three, the Queen extended to the Natives of New Zealand her royal protection and imparted to them all the Rights and Privileges of British Subjects.

Article Two of the Treaty reflected the position that, under British Common Law, a change of sovereignty did not displace pre-existing tribal property rights in the form of aboriginal title over land, forests, fisheries and other resources. Aboriginal rights to land were practically recognised in that the Crown reserved to itself rights of pre-emption over the purchase of land from Maori.

With regard to fisheries, the philosophies and concepts of Maori and British traditions were in far greater conflict. Essentially, the Maori position was that a similar framework of underlying ownership and control applied to both fisheries resources and land. In contrast, the British position with respect to fisheries was inherited from Roman law that held that the ocean was common to all and not susceptible to possession or ownership like land. Accordingly, the English view was that everyone is entitled to fish in the sea and fish were owned by no-one until they were caught at which time they became the exclusive property of the harvester.

Although the legal status of the Treaty of Waitangi has been contentious, the articles of the Treaty provide an excellent framework for discussions about fisheries governance in New Zealand. In particular, the Treaty demands a classification of all fisheries use and management arrangements into those governed by either Article 2 (private property rights) or by Article 3 (government regulation). The articles of the Treaty therefore provide a helpful delineation between the respective realms of fisheries governance and fisheries management.

## British Colonial Fisheries Governance

At the signing of the Treaty of Waitangi, New Zealand formally became a British Colony as a dependency of New South Wales. Towards the end of 1840, New Zealand became a separate colony and a series of governors were appointed through which British rule was effected. In 1852 the British Parliament passed the New Zealand Constitution Act that established an elected House of Representatives, a Legislative Council and a system of elected provincial governments. These provincial governments were dis-established in 1876 and the Legislative Council was abolished in 1950. Since then New Zealand has had a unicameral system of government where power is concentrated in the Parliament.

It is somewhat difficult to pinpoint the exact end of the colonial governance era in New Zealand. It was not until 1947 that New Zealand adopted the Statute of Westminster which confirmed that the New Zealand Parliament alone had the power to make laws for the country, even though that statute had been passed by the British Parliament sixteen years earlier.[3] For practical purposes however, the New Zealand Government exercised control over all domestic matters including fisheries governance from the mid 1860s onwards. In fact, the first piece of fisheries legislation in New Zealand was the Oyster Fisheries Act 1866. Clearly, oyster fisheries were the source of disputes as further legislation relating to them was enacted in 1869 and 1892. This legislation regulated exploitation and specified exclusive rights, including discovery rights for various terms.

The first comprehensive New Zealand fisheries legislation was the Fish Protection Act 1877 followed by the Sea-fisheries Act 1894. The purpose of this legislation was to regulate commercial exploitation of fisheries by non-Maori and these statutes were quite modern in their concern for conservation of fisheries and also for the use of regulation to govern size, quantity, time, place and method of catch. Fishermen were to be controlled through permitting and licensing. While over-exploitation was a concern, evidently under-exploitation was as well. The Fisheries Encouragement Act 1885 was intended to encourage commercial fishing (by non-Maori) generally through export bonus payments. The pattern for New Zealand Government involvement in fisheries governance for the approaching twentieth century was thereby established: the simultaneous encouragement and discouragement of fishing.

By the time all enactments relating to fisheries were consolidated into the Fisheries Act 1908, a comprehensive set of regulatory powers had been accumulated along with an associated Ministerial post (Minister of Marine) and attendant bureaucracy (The Marine Department). Regulatory powers were supported by stringent enforcement measures and applied to commercial and recreational fisheries. Section 77 of the 1908 Fisheries Act stated that 'nothing in this Act shall affect any existing Maori fishing rights' and with a slight change this clause was transmitted into section 88(2) of the Fisheries Act 1983 which stated 'nothing in this Act shall affect any Maori fishing rights'.

In all there were over 15 pieces of sea fisheries legislation which purported to protect Maori fisheries rights in one form or another from the 1860s to the present day. However, New Zealand legislators operated from two assumptions which were consistently rejected by Maori. These were that Maori fishing rights were for non-commercial activity alone i.e. subsistence fishing; and Maori fishing rights were so insignificant in scale that all commercial and recreational use provided for did not entail any disturbance to them.

The consequence of this approach was to erode traditional Maori conservation measures and to force assimilation into the general regulatory framework for commercial fishing (permitting and licensing) by those Maori who wished to pursue commercial fishing as a livelihood. Maori were therefore effectively disenfranchised of their property rights under Article Two of the Treaty of Waitangi but could obtain regulatory consents to fish from the government under Article Three. This was consistent with general government policies to encourage assimilation.

## Widening Horizons: From Three Miles to Two Hundred Miles

By the early years of the 20th century large areas of New Zealand had been deforested to establish pastoral agriculture. In fact, deforestation was so extensive that the government established the Forest Service in 1921 to commence a programme of re-afforestation with exotic tree species to meet the anticipated future timber requirements of the nation. This experience of natural resource depletion permeated government attitudes towards fisheries use, manifest as a strong conservation ethic and tight regulatory control of commercial fishing. These priorities were described by Mr. A.E. Hefford (Chief Inspector of Fisheries from 1926 to 1946) as follows

> *The question of the day as to whether the fishing industry in New Zealand is making as much profit out of fish as it might do is of less urgent importance from the point of view of national economics than the question as to whether we are not already drawing too heavily on our sea fishing resources to be detrimental of their future well being.*[4]

By the time that those remarks were made in 1935, New Zealand was in the depths of the great depression, export markets had largely disappeared and the government was more concerned about ensuring a cheap source of domestic protein for a population enduring high unemployment and suppressed incomes. Accordingly, exports were discouraged, domestic consumption was encouraged and the catching and wholesaling of fish became a small, highly regulated domestic activity. This remained the situation until the 1960s. Constricting fisheries focus onto the small domestic market was also recognised as avoiding the dangers of over-capitalisation and over-fishing attributed to export driven exploitation. The potential of off-shore fisheries was officially discounted.

This heavy-handed regulation of commercial fishing and fish marketing was successful in conservation terms but overlooked the important fact that the jurisdiction of the Marine Department only extended as far as the three mile limit of the Territorial Sea. This excluded large areas of the New Zealand continental shelf. The fishing grounds beyond the Territorial Sea, however, did not escape the notice of nations with distant water fishing fleets, especially Japan. These fleets quickly demonstrated that significant hoki, orange roughy and squid resources were present.

The appearance of foreign fishing fleets operating so close to New Zealand awoke a combined awareness of the economic opportunity represented by nearby fishing grounds and the sense that New Zealand should be taking advantage of those opportunities. In response, New Zealand delicensed commercial coastal fisheries in 1964 and passed the Territorial Sea and Fishing Zone Act 1965 that established an exclusive fishing zone between the three mile Territorial Sea and the twelve mile limit. In addition, development initiatives were introduced, notably government loans and mortgage guarantees to encourage New Zealanders to invest in larger vessels, cool stores and the exporting of fish.

The expansion of the industry was initially slow, but gained pace with the passage of the Territorial Sea and Economic Zone Act 1977 that established an exclusive economic zone of 200 miles (as provided for when UNCLOS was eventually finalised in 1982) and extended the Territorial Sea to twelve miles. Almost immediately, concerns emerged that the economic stimulus was overheating harvesting effort and licensing was re-introduced in 1977. One problem that had become evident was that larger vessels, financed ostensibly to displace foreign fishing effort, were also fishing the inshore grounds of the 'traditional' fleet servicing the local market.

By the early 1980s, there was widespread concern about the decline of inshore fin-fish stocks, particularly the most valuable species such as snapper. As a holding action, a moratorium on the issue of all further fishing permits was introduced in 1982. The general perception was that strong regulatory measures were necessary if the alarming estimates of necessary effort reduction made by the Ministry of Agriculture and Fisheries and National Fisheries Management Advisory Committee (NAFMAC) were to be achieved.

It was in this fraught context that the first major review of fisheries legislation since 1908 was carried out leading to the Fisheries Act 1983. This Act retained the full gamut of regulatory powers accumulated to date but attempted to discipline their *ad hoc* use with an overlay of regional fisheries management plans under which integrated bundles of regulations would be passed and revised. An associated re-organisation of the Ministry of Agriculture and Fisheries was carried out to establish regional offices responsible for the preparation of fisheries management plans.

One immediate effect of the 1983 Act was the cancellation of all unused or 'part-time' commercial fisheries permits. As a consequence 2,260 licences were cancelled without compensation. A significant proportion of those licences were held by Maori fishers for whom the activity of fishing was only part of their traditional livelihood and lifestyle. This permit cancellation generated great bitterness without a large reduction in fishing effort. Furthermore, the slowly grinding mills of fisheries planning established in 1983 could not possibly address the immediate crisis within inshore fisheries.

In 1984, opinions were sharply divided over the appropriate short term measures to halt and reverse over-fishing. There was ample evidence that existing piecemeal effort controls were generally no match for the creative abilities of New Zealand fishers to reconfigure their fishing technology so as to maintain legally high catches. Recognising that heavy regulation was not working, many officials favoured the use of even more extreme 'controlled fishery' powers. At the same time factions within the industry gained NAFMAC support for government compensation of remaining 'full-time' fishermen affected by any new effort reduction or industry restructuring measures. Examination by officials of boat buy back schemes elsewhere only added to government alarm about the financial implications of this prospect. The third option mooted to address overfishing was to establish sustainable Total Allowable Catches (TACs) and allocate Individual Transferable Quotas (ITQs) to incumbent fishers.

## Introducing the Quota Management System (QMS)

The expansion of New Zealand fisheries jurisdiction and the 'New Zealandisation' of fishing within its EEZ industry led with remarkable rapidity to a fisheries management crisis. This was all the more bewildering to the government given the existence of a well trained and dedicated bureaucracy, fisheries scientists, enforcement capability and the availability of comprehensive regulatory powers. Indeed, by the 1980s New Zealand had approximately 4,000 individual fisheries regulations. At the same time, New Zealand recognised that fisheries management was not just a problem of suppressing fishing but that it needed to extract maximum value from fisheries resources.

During the 1970s, the wider New Zealand economy was stagnant and living standards were falling relative to many other countries. The 'oil shocks' of 1973 and 1979 were keenly felt but the more profound impact was from the changing relationship between the UK and its former colony. New Zealand's prosperity had been underpinned by a simple mutual relationship wherein New Zealand supplied Britain protein and Britain supplied New Zealand an array of manufactured goods. Notwithstanding the fact that the UK had signalled its intention to join the European Union as early as 1961, the entry of the UK into the EU in 1973 was still accompanied by successful pleadings by New Zealand for special trading arrangements to soften this blow to the New Zealand economy. The last New Zealand cheese exports to the UK under these arrangements occurred in 1978.

New Zealand had, in fact, already made somewhat successful efforts to diversify its export markets, if not its economy, during this period. The share of New Zealand exports accounted for by the UK dropped from 50 per cent in 1965 to 7 per cent in 1989 (Belich, 2001). Fisheries development held the promise of supporting a diversification of the New Zealand economy away from one that supplied a narrow range of pastoral farming commodities optimised for British tastes. This promise would not be realised if the fisheries themselves were depleted. However, the fisheries management crisis of the early 1980s was a tiny element of the national economic, financial and economic crisis that led to an early general election in July 1984.

In 1984 the incoming Labour government embarked upon a sweeping set of reforms including financial deregulation, the phasing out of agricultural and consumer subsidies, (including subsidies to the seafood sector) and corporatisation or privatisation of state owned and operated businesses including the railways, the Bank of New Zealand and Air New Zealand. The general thrust of these reforms was to make greater use of market forces in determining the shape of the economy. It was in this environment of comprehensive and bold liberalising reform that the most radical response proposed to address the overfishing problems of the day (introduction of ITQs) was adopted.

The details of what became known as the Quota Management System (QMS) were consulted upon in 1984/85. ITQ was to be denominated as a right to catch a specified tonnage of a particular fish stock each year. Licensed fishermen were to be given provisional ITQ based on their catch history records and the government offered to buy sufficient of these rights to reduce catches to sustainable levels. If insufficient willing sellers emerged, ITQ rights would be proportionately reduced to fit the Total Allowable Commercial Catch (TACC) desired. The quid pro quo was that if TACCs could be increased in the future, the government would be entitled to sell any ITQ rights made available as a result. In the end,

15,800 tonnes of quota reductions were achieved by purchase and 5,700 tonnes of quota reductions were achieved by administrative reduction (Clark, Major and Mollet, 1989). The cost of these reductions was approximately $42m but government revenues from quota sales and leases were approximately $96m by the end of 1987.

Commercial fishers were divided on the subject of ITQ. Share-fishers who did not hold licences themselves were opposed as they would receive no ITQ. Former part time fishers who had lost licences and associated catch history in 1983 saw themselves as victims of a Machiavellian plot to deny them quota (their grievance had substance but their conspiracy theories did not). Many incumbent fishers worried that their reduced catching rights would leave them with unviable fishing plans for their vessels. However, even though many practical details were still unclear, most recognised that government funding of effort reduction was contingent upon acceptance of ITQ. Rock lobster fishers remained resolutely opposed to ITQ and that fishery was excluded from the 21 inshore species and 8 deepwater species introduced into the QMS on 1 October 1986. ITQ was a perpetual and tradable right but ITQ owners were obliged to pay resource rentals to the government.

It is hard to avoid the conclusion that neither supporters nor opponents of the QMS fully understood the ramifications of ITQ and that their individual understandings were also plainly incompatible. The diverse justifications for imposing resource rentals on ITQ provide a good illustration. Resource rents were supposed to be struck at levels that would remove any super-profits from fishing. The rents would provide revenue to offset fisheries management costs incurred by the government. They would also provide a return to the public (addressing legally misconceived complaints about the sale of a publicly owned asset). Achieving low ITQ values would make the purchase of quota affordable in the event that the government was required to buy and retire ITQ for sustainability reasons. Low ITQ values would also discourage speculative ownership of quotas and minimise entry costs to the industry.

These arguments generally contradict the economic rationale for ITQ advanced by writers such as Gordon (1954) and Scott (1998). This rationale emphasises the latent power of incentives for conservation and wise use if individual resource users have secure rights that allow them to capture the capitalised benefits of resource conservation, cost minimisation and value maximisation. Clearly, those incentives require the possibility of high future quota values and the capture of that value by quota owners. There is no strong evidence from the time that the government or officials generally regarded the introduction of the QMS as a fundamental change to the role of the state in New Zealand fisheries governance. They certainly did not contemplate that they were exposing the government to Maori claims that would cost taxpayers $280m to settle.

## Worlds Collide: The Maori Injunction and Fisheries Settlement

With the introduction of the QMS, Maori immediately organised legal action to injunct expansion of the QMS, arguing that the allocation of ITQ conflicted with and undermined existing Maori fisheries rights and was an action inconsistent with the fiduciary duty of the government under the Treaty of Waitangi to secure and guarantee those fisheries rights mentioned in the English version of Article 2. The injunction was granted in 1987 by the High Court and various Maori plaintiffs began organising themselves for a substantive legal hearing on the issue that would consider the nature and extent of those rights and the extent of harm to those rights caused by recent government policies and actions.

The success of the Maori injunction should probably have come as no surprise to the government given two separate recent signals it had received from the Courts and the Waitangi Tribunal. The Waitangi Tribunal had been established by Act of Parliament in 1975 to investigate Maori claims about breaches of the Treaty and to make recommendations to the government about remedies for any such breaches. Large Maori protests about historical land confiscations such as the Land March (*hikoi*) of 1975 and the occupation of Bastion Point in Auckland in 1977/78 had confronted non-Maori with the reality of a resurgent Maori culture that rejected complete integration. In 1985, the Tribunal obtained a remit to hear historical claims, one of which was a claim by Muriwhenua tribes to fisheries. The Tribunal recommended to the government that the introduction of the QMS should be delayed until that investigation was completed (not until 1988). That recommendation was rejected.

The second signal was the result of Court case of a Maori (Tom Te Weehi) charged with an offence under 8(1)(b) of the Fisheries (Amateur Fishing) regulations 1983 (Scott 1998). Mr. Te Weehi offered the defense that he was not engaged in recreational fishing but was exercising a Maori fishing right under the authority of an elder (*kaumatua*) of the local tribe (*Ngai Tahu*). The Court agreed and on the basis that section 88(2) of the Fisheries Act 1983 stated that 'nothing in this Act shall affect any Maori fishing rights' the case was quashed. The government did not appeal this judgment which was the first time that customary fishing rights had been acknowledged by a Court, the implication being that such rights continued to exist until they were expressly taken away with the right holder's agreement.

It is perhaps understandable that the government had little evident appetite for the substantive proceedings that would follow the injunction and invited Maori to select representatives to negotiate a settlement over the issue. By 1988, a team of Maori negotiators was appointed with a mandate confirmed at a national *hui* (meeting) to seek a settlement of no less than 50 per cent of ITQ. No general agreement between the negotiators and government was achieved but an interim arrangement (pending full settlement) was passed into law by the Maori Fisheries Act 1989.

The interim arrangement provided for the establishment of the Maori Fisheries Commission (MFC) with $10m of funds. The government undertook to buy 10 per cent of the ITQ in each fish stock on a willing buyer-willing seller basis and transfer it to the MFC. The Act also provided for the establishment of *taiapure* areas in coastal waters intended to make better provision for Maori fishing rights. This intention has not been realised as in practice this measure amounts to a limited ability of Maori to influence government regulations in particular localities rather than exclusive rights to exercise *rangatiratanga*.

The acquisition of 10 per cent of ITQ took four years and cost the government approximately $120m. As quota was progressively transferred to the MFC it was leased out under arrangements where some preference was given to Maori lessees. Profits from leasing were used by the MFC to buy other fisheries assets. By the time these interim arrangements reached their end, New Zealand had five years of experience with the QMS and its unrealised potential was starting to be recognised. A Task Force established to advise the Minister of Fisheries on appropriate legislative change and reform found:

The QMS which is the main device for managing New Zealand's fisheries is a suitable foundation for the development of a consistent and comprehensive fisheries management regime.

- The principles of the QMS do not appear fundamentally at odds with the Treaty of Waitangi. Indeed there appears to be scope to adapt the QMS as a means of providing effective recognition of Maori fishing rights secured by the Treaty.
- The QMS is consistent with sound environmental management. Management systems which incorporate individual tradable harvesting rights in defined fisheries do not preclude sensitivity towards wider ecosystem effects of fishing activity.
- A fisheries management regime based around the QMS is capable of protecting the rights of recreational fishers, including a guaranteed continued right to engage in recreational fishing at minimal cost.

Nothing has happened in the intervening 17 years in New Zealand to contradict these basic findings and the QMS received explicit endorsement from Maori in the Deed of Settlement that settled all fisheries claims in 1992. The Deed states '*Maori Endorses the QMS and acknowledges that it is a lawful and appropriate regime for the sustainable management of commercial fishing in New Zealand*'.

That agreement achieved a discontinuance of all Maori proceedings against the government over fisheries. For its part the government provided $150m of funding to allow The Treaty of Waitangi Fisheries Commission (the re-named MFC) to acquire a 50 per cent interest in Sealord (New Zealand's largest fishing company). Furthermore, the government would allocate 20 per cent of ITQ in any new quota species to the Commission for distribution to Maori. The government also undertook to make regulations providing for customary (meaning non-commercial) fishing. Today, under the commercial management of Maori, the $280m of Settlement assets have grown in value to more than $1b.

The Deed was given effect through the Treaty of Waitangi (Fisheries Claims) Settlement Act 1992. The government, tax payers and commercial fishers now had far greater certainty about the extent of Maori rights. However, for Maori an intense debate on the way in which the assets and benefits of the Settlement should be dealt with then occurred. By 1992, the degree of urbanisation of the Maori population was the same as that for non-Maori (over 80 per cent). Most Maori no longer live and work within their tribal areas. While genealogy (*whakapapa*) remains central to Maori identity, *iwi* and *hapu* no longer retained the integrated political/economic/cultural/social functions of 1840. There was a tension between views of the Settlement that emphasised restitution (return assets to *iwi* and *hapu*), and desired Settlement outcomes that emphasised maximising future commercial development capacity or the delivery of assistance to address contemporary Maori socio-economic problems in cities.

## Fisheries Right Holders Organise (1992–1999)

Individual fisheries rights holders have interests in common with all other similar right holders. All fishing activity entails external effects on others and these externalities can potentially be managed through agreed collective arrangements. In 1992, Maori and non-Maori were approaching this organisational challenge from opposite directions. Maori were the recipients of a Settlement whereby the government placed a pool of assets into the Treaty of Waitangi Fisheries Commission on behalf of all possible claimants and beneficiaries. Maori therefore faced a problem of dividing up the Settlement or rights to its benefits. As well as the stark differences in vision over what the Settlement represented or the purposes that Settlement assets might underwrite as described above, there were many structural possibilities within each vision.

It is a remarkable and unheralded achievement that Maoridom was able to reach agreement over the allocation of Settlement assets. This agreement was implemented through the Maori Fisheries Act 2004 and required mutual recognition of contemporary *iwi* (as opposed to *iwi* in 1840) and the establishment of mandated *iwi* organisations to receive Settlement assets. Quota was divided into two categories: inshore and deepwater. Inshore quota was allocated to tribes with coastline adjacent to the fishery and deepwater quota and cash was allocated on the basis of relative tribal populations. Corporate investments, including the shareholding in Sealord, were placed into an *iwi*-owned holding company (Aotearoa Fisheries Limited) where shares were also allocated to the same mandated iwi organizations on a relative population formula. All *iwi* and therefore potentially all Maori are beneficiaries of the Settlement.

Trusts were set up under the auspices of a further reconstituted Treaty of Waitangi Fisheries Commission (Te Ohu Kai Moana Trust Limited) to address general issues of urban Maori, aquaculture and fresh water fisheries which individual mandated *iwi* organisations were considered to be ill equipped to deal with. The negotiation and implementation of the Settlement was testament to the capacity of Maori society to develop new arrangements consistent with both Maori culture and twenty first century commercial realities. By itself, the demonstration of this confidence and capability was an important outcome of the Settlement.

Having resolved questions of ownership, the challenge is now for *iwi* organisations to develop effective means of working together with each other and other quota owners to protect and enhance the value of ITQ. The interests of all ITQ owners are now broadly aligned through the possession of a generic property right. Unfortunately, the same alignment has not been achieved with other fish harvesters. Even the complementary customary fisheries regulations promised by the government under the Settlement have been clumsily handled in that (in the North Island) there is no necessary connection between the Maori mandate and representation processes and structures in place for the commercial fisheries Settlement and those for customary fishing. As a result, demands for customary fishing frequently do not internalise their impact on Maori owned fishing businesses.

While Maori were working through Settlement negotiations with the government and each other, quota owners were exploring the potential of their new property right. Rather than starting with a pool of ITQ, their initial position was one of extreme disaggregation. The potential for ITQ to encourage responsible fishing by right holders was outlined in major reviews of the QMS conducted for the government by Professor Peter Pearse (Pearse, 1991) in 1990/91 and the Legislative Task Force a year later. These emphasised the desirability and opportunity to transfer fisheries management responsibility and cost to ITQ owners acting collectively through proposed quota holders' associations. The incentives on ITQ owners strengthened in 1990 when the government (faced with having to make some large quota purchases to reduce catches in the orange roughy, hoki and lobster fisheries) converted quota from a fixed tonnage right to a proportional right.

As an associated measure, resource rentals were frozen and used to fund compensation of quota owners in fisheries most affected by the cuts. At the end of the compensation arrangement in 1994, resource rents were abolished and ITQ owners became subject to government levies to recover fisheries management costs. By 1994, therefore, ITQ owners realised that they would bear the full financial consequences of adverse fisheries events, but were potentially the beneficiaries of fisheries improvements or enhancements. As funders, they were also keenly interested in the cost efficiency of required fisheries management services but they lacked practical arrangements through which ITQ owners could manage these common risks and opportunities.

The Challenger Scallop Enhancement Company incorporated in 1995 was to provide a model for such arrangements. This scallop fishery had been severely overfished in the 1970s and was completely closed for two years in 1981 and 1982. It was opened as a controlled fishery in 1983 under extremely tight regulatory control including the imposition of a compulsory harvester levy to fund a government run scallop re-seeding and rotational fishing trial. The success of the programme led to a 1989 proposal by licence holders to bring the fishery into the QMS, to make allowance for a Maori quota allocation (prior to the Settlement of 1992) and to support funding of enhancement voluntarily. ITQ was eventually allocated in 1992 and the Company developed an enhancement plan for the fishery supported by an Agreement for Provision of Services between the Company and the Ministry of Fisheries. This was the first example of a right holder fisheries plan with integrated fisheries management service specification, funding and delivery.

Since then, the Company has facilitated the establishment of the Challenger Scallop Recreational Advisory Group and provides management services for associated dredge oyster and some fin fisheries in the area. The impressive track record of the Company was reflected in a 1998 Memorandum of Understanding wherein the government pledged to maximise the ability of stakeholders to act in a collective manner and develop opportunities for self management. Maori have continued to be supportive shareholders of the Company.

The Settlement with Maori, the sustainability improvements under the QMS, the economic success of the industry, positive academic reviews and the emergence of exemplars like the Challenger Scallop Enhancement Company seemed to chart the future direction of fisheries management in New Zealand. The dangers of over-fishing had been averted without choking off the development of a vibrant and unsubsidised export industry by then employing approximately 10,000 people. Indeed, the contribution of the seafood sector to the diversification of the New Zealand economy is confirmed by 2008 seafood exports of $1.3b (nearly 30 per cent more than the combined value of sheep-meat and wool exports for that year) and quota assets estimated to be worth $3.8b by the government.[5]

In this mood of optimism, a new Fisheries Act, drafted in 1996, was the first in New Zealand history to contain a statement of purpose, provision for the preparation and approval of right holder fisheries management plans and the devolution of responsibilities to right holder organisations. The lifting of the injunction against the use of the QMS allowed the remainder of commercial species to be introduced in to the QMS so that today, quota has been issued (by tender in recent years) to 629 fish stocks. At first sight, the foundations for the expansion of a rights-based approach to fisheries management advocated earlier by Pearse and others were in place. However, all pre-existing government regulatory powers were retained in the new Act and the tide of liberalising reform that had been running with declining speed since 1984 was about to turn.

## Rebuilding the Centre (1999–2008)

In 1999, a Labour Government was elected and held office for three terms until October 2008. The public policy position of this government was based upon the proposition that previous reforms (ironically initiated by an earlier Labour government) had gone too far, were inequitable in their economic and social outcomes and had eroded the capacity of the government to deliver the range of services that New Zealanders expected of it. The new suite of policy settings were described as 'rebuilding the centre'. Effectively, this was a reassertion of state power, supported by an expanded public sector, especially in the form of a regulatory bureaucracy.

In fisheries management, the legislative platform for greater central regulation was already in place. While the QMS and ITQ are prominent features of the New Zealand fisheries management regime and the initial 29 species introduced to the QMS in 1986 were the major commercial fisheries, there were still approximately 100 other fish species fished commercially at that time. These continued to be managed under historical licensing and regulatory arrangements. New Zealand had (and continues to have) a dual fisheries management system: ITQ and non-ITQ.

Initially, ITQ was regarded by officials as an adjunct to their traditional powers to control fisheries use through central government regulation. As the attributes of ITQ were strengthened over time and as its status grew as the 'currency' of the Maori fisheries Settlement, these very attributes came to be seen as possible threats to those powers and their associated perquisites. As a result, the encroachment of right holders into the fisheries management space, or tier, through the processes of devolution and management planning initiatives was halted in 1999. The government declared it would 'take the lead in fisheries management planning'. No plans have yet been completed. The devolution of suitable fisheries management services was also halted, the last being devolution of registry services in 1999 to an industry owned company 'Fishserve'. The performance of Fishserve is generally acknowledged by both the government and the industry to be accurate, transparent and cost efficient.

Between 1999 and 2008, staff numbers and budget for the Ministry of Fisheries both increased by approximately 100 per cent. In its traditional areas of regulation, the Ministry of Fisheries has been very active. A senior official recently estimated that there were perhaps now 8,000 fisheries regulations of one kind or another.[6] This substantial investment in personnel and regulations has re-established government command and control over fisheries but at the cost of management sclerosis. Fisheries in 2009 are characterised by growing sectoral conflict and a frustration at barriers to innovation and development.

## The Future: Just Reconciliation or Just Wreckage?

Attempts by recent governments to reconcile national fisheries governance responsibilities with duties to protect Maori fishing rights secured by the Treaty of Waitangi have been uncomfortable. The Te Weehi case confirmed the existence of private and perpetual rights in fisheries and the historical failure of the government to recognise and protect those rights, blinded as it was by cultural assumptions about the appropriate legal framework for fisheries. The acceptability of ITQ to Maori as compensation for fisheries claims was due to its attributes as a secure perpetual property right. All ITQ, whether owned by Maori or non-Maori therefore now has that character. And the government cannot undermine ITQ without undermining the integrity of the Settlement. Given the relatively unbridled powers of the New Zealand Parliament, it is quite possible to undermine quota under the cover of statute. However, actions inconsistent with the simple framework of the Treaty of Waitangi are readily seen for what they are by voters, including the 15 per cent of New Zealanders of Maori descent.

The recognition of ITQ and Maori customary rights immediately raises questions about the responsibilities of other fisheries users to those right holders and the nature and extent of recreational and aquacultural rights. While the questions may be obvious, they are currently unanswered. The legal framework for aquaculture, for instance, tries to pretend that the expropriation of space from quota management areas to establish aquaculture management areas is not an issue of property rights but that the loss of space is an 'environmental effect'. Similarly, other spatial expropriations are generally not compensated because, individually, their effects are not considered to be 'undue'.

This sort of pretence is ultimately unsustainable. New Zealand has been able to face up to the concept of bi-culturalism but while two cultures can co-exist, two conflicting fisheries management philosophies with attendant regimes cannot. The reconciliation achieved by the Settlement between the role of government and Maori fishing rights has simultaneously removed conflict between Maori and non-Maori commercial fishers. This is a powerful demonstration of the positive effects of allowing people to interact directly with each other under the framework of secure tradable property rights. These effects were not intended by the government when the QMS was first introduced, neither was ITQ intended to acquire the attributes it has today. But the worth of government policies, including policies that shape fisheries governance should surely be judged by their outcomes, rather than by their intentions.

## Relevance to the Commonwealth

The New Zealand experience is not unique. Schizophrenic policies of fisheries expansion and constraint feature in many, if not all, fisheries management regimes at some stage in their history and these tensions are particularly acute in developing country economies. Maori traditional systems for fisheries management also find their equivalent in cultures throughout the world. The re-incorporation of some of these Maori elements into the QMS has been crucial to its success. It should come as no surprise that systems that have succeeded and endured for long enough to become 'traditional' contain features of merit that are still relevant today.

What is perhaps less well recognised elsewhere is the role that property rights can play in countries with a colonial heritage that are faced with addressing the problems of fisheries management. The case of New Zealand illustrates the subtle ways in which unexamined colonial assumptions can constrain contemporary policy and structural choices. The New Zealand experience is also that property rights must not only be secured and guaranteed by the state, they must be enforceable against the state (otherwise they are not rights). Private rights and government rights are, in this respect, competitors for the same space. It therefore needs to be acknowledged that there is a natural propensity for government to re-assert control and that right holders must exercise a constant vigilance for signs of this propensity. One such sign is a characterisation by officials of rights-based systems as instruments of government.

New Zealand is one of the least corrupt nations on earth. Recognition of, and respect for, property (real or personal) is a foundation for building a corruption free society. The history of fisheries governance in New Zealand demonstrates that property rights rooted in pre-colonial culture can be given expression in modern fisheries management and that such expression has not only helped address the tragedy of the colonial commons at sea but has also contributed to a far wider social and economic enrichment of New Zealand. As a Chair of the Treaty of Waitangi Fisheries Commission has stated:

> *When we had the commons out there, you went out and the fish belonged to no one until they were caught. The whole model of the commons, from our point of view, was a device, driven by the majority culture, for preventing particular Treaty property rights being given effect to. It was thus a great irony when this country moved towards a model of property rights in fisheries under a quota management system. Although it was a model driven by a whole lot of notions, principally sustainability, it was very much centred on notions of property as far as the fishers themselves were con-*

> *cerned. Even though the quota management system was not invented, developed and conceived to give effect to Treaty rights it provided, conceptually the key to that historical resolution. In New Zealand history that may yet stand as its greatest triumph.* (O'Regan, 1996)

The QMS also provides, conceptually the key to the resolution of conflicts between commercial fishing, customary fishing, recreational fishing, aquaculture and other marine activities that require the exclusive occupation of space. That potential is to expand the allocation of well defined rights and compatible rights to those sectors. Through the direct involvement of self interested right holders in fisheries management, the QMS also provides, conceptually the key to the evolution of a dynamic and sophisticated approach to fisheries use and conservation. Only time will tell whether New Zealand can complete the process of translating these concepts and opportunities into a reality.

## References

Belich, James (2001) *Paradise Reforged* (Auckland, London: Allen Lane).

Clark, I.N., Major, P. and Mollett, N. (1989). 'The Development and Implementation of New Zealand's ITQ Management System' In *Rights Based Fishing*, edited by Philip A. Neher, Ragnar Arnason and Nina Mollet (Dordrecht, London, Kluwer Academic).

Gordon, H Scott (1954) The Economic Theory of a Common Property Resource: The Fishery. *Journal of Political Economy* 62, pp. 124–42.

O'Regan, Sir Tipene (1996) Video interview with Jerry Morris, shown at Fishing Industry Association Conference, April 11.

Pearse, Peter H. (1991) *Building on Progress: Fisheries Policy Development in New Zealand*.

Scott, Anthony (1998) Development of Property in the Fishery, *Marine Resource Economics*, 5: 289–311.

## Notes

1 Ngai Tahu Maori Trust Board v Attorney-General and Ors, 2 November 1987, (CP 559/87).
2 Treaty of Waitangi, Article 2, English version.
3 Wilson, John, 'Government and Nation' Te Ara – the Encyclopaedia of New Zealand.
4 Appendix to the *Journals of the House of Representatives*, vol. 3, H.15: 14 (1935)
5 Statistics New Zealand 2009 'Fish Monetary Stock Accounts: 1996–2008', Wellington, Statistics New Zealand.
6 Crothers G.S. Deputy Chief Executive, Ministry of Fisheries pers. comm. 2008.

# 9 Enough to Eat? Fisheries and Food Security

Tim Bostock and Suzannah Walmsley

Fisheries contribute to food security in two main ways. First they provide a high quality food containing a range of nutrients essential to a healthy diet and particularly relevant to maternal and child development. Second, they generate wealth that provides livelihoods and that can fuel economic growth. Being a *renewable* natural resource, these benefits may be generated indefinitely and in many cases enhanced. This chapter examines some of the challenges to achieving this ultimate goal.

## Introduction

Fish is the main source of animal protein for about one billion people around the world and provides a fifth of total animal protein nutrition within low income food deficient countries. *Per capita* consumption varies considerably from place to place. Current estimates of global consumption are 16.7 kg (FAO, 2009) while in some places (Brazilian Amazon; Sierra Leone) consumption can reach 150 kg per person even amongst the rural poor. In addition to protein, fish provides nutrients required for a healthy diet, particularly for maternal and child development. These include essential polyunsaturated fats, vitamins, minerals and elements such as iodine.

Fisheries also contribute to the economies of many poor countries through providing jobs and generating wealth. One third of global fish production enters international trade. Of this, fishery products top the list of net exports of selected agricultural commodities by developing countries.

Yet, the sustainability of these benefits is now under threat. With a few notable exceptions, fisheries governance globally has been ineffective in ensuring fisheries are exploited in ways that can generate food and wealth in a sustainable manner. Overexploitation, illegal fishing and conflicting demands for marine resources and access, all have negative impacts on fisheries' contribution to food security. Many fisheries around the world are now exploited well beyond biologically sustainable limits because of the inability of policies to address the fundamental problem of open access. Policy objectives are often ill-directed. Perverse incentive structures remain that perpetuate inefficiencies within the industry. The current cost of this gross inefficiency both to the global economy and to the people that depend on fisheries for work and food, is likely to amount to some US$50 billion per year (World Bank and FAO, 2009).

The extent to which fisheries are able to operate on a sustainable basis, producing wealth and social benefits, is directly linked to the management arrangements that govern fisheries. The challenge for capture fisheries is to put in place economically rational management systems backed by supporting institutions to ensure the sustainability of fisheries in the years ahead. Such systems will also lead to enhanced resource resilience in the face of external shocks such as climate change.

Aquaculture (fish farming) has a similar key role to play in food security, both through its contribution to local and international food supplies, and to economic growth. Aquaculture offers notable efficiencies when compared with traditional livestock agriculture systems and represents a significant growth opportunity. However, its development must be managed in an environmentally-sensitive manner, and it must be considered alongside capture fisheries rather than a replacement for it, given that capture fisheries are substantially more important in global economic terms.

This chapter outlines the concept of food security, briefly reviews fisheries and aquaculture production and then outlines the ways in which they contribute to food security. The factors that threaten fisheries' and aquaculture's contribution to food security are discussed, and means by which their contribution can be maintained and enhanced are highlighted.

## What is Food Security?

Food matters to us all. Food security stands at the intersection of many disciplines and is a complex issue. Governments everywhere have a responsibility to ensure that everyone has enough to eat – food needs to be available, but it also needs to be affordable and accessible through a resilient and reliable supply system. Worldwide around 852 million people are chronically hungry due to extreme poverty, while up to 2 billion people lack food security intermittently due to varying degrees of poverty (FAO, 2003a).

FAO (2002) notes that *"food security exists when all people, at all times, have physical and economic access to sufficient, safe and nutritious food to meet their dietary needs and food preferences for an active and healthy life"*. As such, it does not just concern food production; it also incorporates the physical and economic access to sufficient safe and nutritious food to meet dietary needs.[1]

## Fisheries Production and Trade

### 1. Status and production

At the opening of the London Fisheries Exhibition in 1883, T.H. Huxley confidently remarked that important sea fisheries were *inexhaustible* under the then present modes of fishing. He based this conviction on

> *two grounds – first that the number we catch is relatively insignificant; and secondly that the magnitude of the destructive agencies at work upon them is so prodigious that the destruction effected by the fisherman cannot seriously increase the death rate …*

Indeed, such 'bounty' of the seas was a widely accepted notion until relatively recently.

This confidence was short-lived. By around the middle of the 20th century, the impacts of an expanding mechanised fishing fleet were well recognised. It had already become apparent that fishing productivity could be increased through reductions in fleet catching capacity. Yet throughout the latter half of that century, many countries strove to increase fleets and production.

Driven by increasing technology and ever more powerful fishing vessels, fish catches correspondingly rose during the 1980s and 1990s. But since then production from capture fisheries has stagnated, despite increasing effort, reaching some 92 million tonnes in 2006 (including marine capture and inland capture) (FAO, 2009).

By the 1990s, data on overfishing was increasingly available. It was becoming clear that many of the world's fish stocks were in difficulties. The latest edition of the *State of World Fisheries and Aquaculture* (SOFIA) (FAO, 2009), underscores the significance of these difficulties. Data for 2007 showed that 28 per cent of the world's fish resources were 'over-exploited, depleted and recovering', 52 per cent 'fully exploited' and 20 per cent 'underexploited and moderately exploited'. The corresponding data for 1974 were 10 per cent, 50 per cent and 40 per cent. Aside from the general concern about under-reporting of FAO data, it is further depressing to note that the situation reported in 2009 is actually worse than that reported in the previous SOFIA, where the corresponding percentages were 25 per cent, 50 per cent and 25 per cent. What is clear is that despite some reduction in fishing pressure, there is no sign of global improvement in the state of fish stocks.

Of the 92 million tonnes of global production, marine capture fisheries currently account for 82 million tonnes, and the remaining 10 million tonnes comes from inland capture fisheries (although data on the latter are most likely heavily underestimated due to un- or under-reported catches). Aquaculture production reached 51.7 million tonnes in 2006, with a value of US$78.8 billion (of which 70 per cent and 80 per cent, respectively, is accounted for by China). The contribution of aquaculture to global supplies of fish (and crustaceans, molluscs and other aquatic animals) has increased from 3.9 per cent of total production in 1970 to 36.0 per cent in 2006 (by weight). In 2006, 47 per cent of the world's fish food supply came from aquaculture. Aquaculture continues to grow more rapidly than other animal food-producing sectors, maintaining an average annual growth rate of almost 9 per cent worldwide since 1970 (6.5 per cent excluding China). The growth rate from 2004–2006 was 6.1 per cent in volume terms and 11.0 per cent in value terms.

Some people believe that with the stagnation in capture fisheries production, aquaculture is the only potential source of a future increase in the supply of fish. However, whilst aquaculture has significant potential, it needs to be considered alongside, not as a replacement for, capture fisheries, which are substantially more important than aquaculture in global economic terms.

## 2. Trade and consumption

Fish and fisheries products are one of the most widely-traded commodities. For developing countries, fish traded internationally represents by far the most valuable of agricultural commodities (Figure 1). The percentage of fisheries and aquaculture production currently entering international trade is about 38. The value of this has increased from US$8 billion in 1976 to US$78 billion in 2005. Over half of this trade originates in developing countries and most is destined for the developed markets of the EU, USA and Japan. Regional trade in a wide range of fisheries products, both from capture and farmed origins, is also particularly important. For example, intra-African trade of fish is considerable, absorbing up to 50 per cent of fish catches in some inland fisheries and valued conservatively at around US$284 million per year.

**Figure 1: Global fisheries trade**

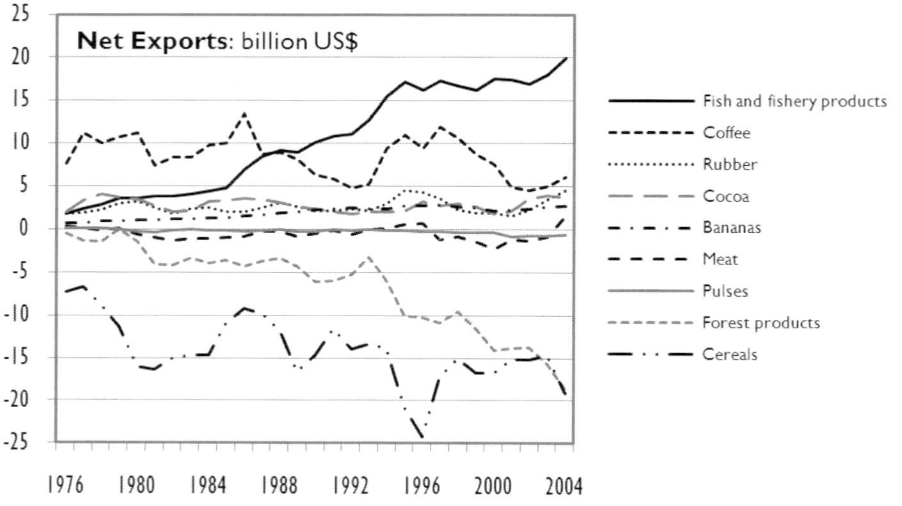


Net export revenues from fisheries, agricultural and forest commodities for developing countries, 1976–2004. One third of global fish production enters international trade. Of this, fishery products are among the most traded commodities and top the list of net exports of selected agricultural commodities by developing countries. Source: DFID & MRAG (2008). Data source: Fisheries – FAO FISHSTAT from Fisheries and Aquaculture Information and Statistics Service; Agriculture and Forestry – FAOSTAT.

Current (FAO, 2009) estimates for per capita fish consumption indicate a slight increase in global supplies to about 16.7 kg in 2006 from an average of 9.9 kg in the 1960s. Corresponding per capita consumption figures for Oceania, North America, Europe, Central America and the Caribbean, and South America are 24.5, 24.1, 20.8, 9.5 and 8.4 kg, respectively. In some places such as the Brazilian Amazon and Sierra Leone, average annual fish consumption ranges between 50 kg and 150 kg per person amongst certain sectors of the rural population. In developed countries average annual fish consumption can be relatively high. For example, 24.3 kg in the USA, 55.4 kg in Portugal and 63.2 kg in Japan (FAO, 2006b).

## How Fish Contribute to Food Security

Fisheries are an important part of food security, particularly for many poor people in developing countries. Fish is the main source of animal protein to about one billion people globally and also provides essential nutrients – minerals, iron and oils. In low income food deficient countries (LIFDCs), they make up 22 per cent of animal protein consumption overall and in Asia, this rises to 26 per cent (FAO, 2003a). In coastal areas and around major river systems the dependence on fish is usually higher (FAO, 2005b). The importance of small-scale fisheries in particular for food security is emphasised by FAO (FAO, 2003b). In addition to providing a food source, fisheries contribute to livelihood security and sometimes act as social 'safety nets'.

### 1. Interpreting the data

Despite a wealth of evidence, the extent to which fisheries contribute to food security is poorly recognised. In part this is because of a lack of monitoring and production information, especially from small-scale fisheries which are the source of much of the fish consumed by the poor in developing countries. This is well illustrated in the case study of Mozambique (Box 1). One study estimated that small-scale fisheries may account for between 25 per cent and 33 per cent of worldwide capture fisheries production (Chuenpagdee *et al.*, 2006).

Statistics on the proportion of animal protein from fish in people's diets often appear to show that fish make up only a small part of the diet of people in many countries around the world. However, in many cases, the consumption of fish is much greater than the statistics indicate. There are two main reasons for this.

Firstly, the statistics are calculated on a national basis, giving an average figure per head of population, without taking into account regional differences. For example, coastal and riverine communities may consume far larger amounts of fish than the average (for example, see data for Brazil and Sierra Leone above).

Secondly, the statistics are calculated based on a mass balance model, taking the total fisheries production of the country, plus fish imports, less fish exports, divided by the population size. In many cases, particularly in developing countries, there is a lack of resources for the collection and analysis of data on fisheries catches. This is exacerbated for the small-scale sector, which by its nature tends to be dispersed, with a multitude of landing sites for many small vessels and gears, making the collection of catch data very difficult. This results in an underestimation – sometimes severe – of the volume of small-scale fisheries production. This is particularly an issue for fish consumption by the poorer segments of the population in developing countries, as most of the fish they consume originates from such small-scale fisheries. Hence, an underestimation of small-scale production can result in an underestimation of the fish consumption per capita at national level.

## Box 1: Mozambique – assessing the real contribution of fisheries to food security

Mozambique provides an illustrative example of the lack of recognition of the importance of fish in people's diets as a result of the underestimation of small-scale fisheries catches. Mozambique has 70,000–100,000 artisanal fishers, dispersed along a coastline of 2,700 km and several inland water bodies (Lake Cahora Bassa and Lake Niassa). There are also semi-industrial and industrial fisheries, based on shallow-water shrimp, and foreign vessels targeting tuna and other large pelagics. The sampling system for collecting fisheries statistics, run by the Directorate of Fisheries Administration, was developed based on the semi-industrial and industrial fisheries, and thus is focused on the main fishing ports.

On the rare occasions that small-scale fishers landed their catch at the semi-industrial and industrial ports, the data collectors would register their catches. The annual artisanal catch was then calculated as the sum of these sampled catches (without taking into account days, locations and vessels that had not been sampled) (Walmsley *et al.*, 2002). It was recognised that most small-scale fishers land their catches at other locations – the over 800 artisanal landing sites (mainly beaches but also inland sites) – and both the Fisheries Research Institute (IIP) and the Small-scale Fisheries Development Institute (IDPPE) had developed their own catch sampling systems better-adapted to the characteristics of the small-scale sector. Nevertheless, the data from the Directorate of Fisheries Administration was used in the national production statistics. As such, the small-scale sector was estimated to contribute around 8,000 tonnes annually to an overall production of around 30,000 tonnes (Ministry of Fisheries).

In 2002, the fisheries institutions began a process of analysing the various systems to collect small-scale fisheries statistics, and identified that the most appropriate system was the one being implemented by IIP – a system based on random stratified sampling of landings of the small-scale landing sites, followed by extrapolation to account for non-sampled active vessels and non-sampled days and landing sites. Great efforts were made to expand this system to cover a larger proportion of the coastline, so that it could be adopted as the official system for statistics on capture production from the small-scale sector. As a result, the estimate of small-scale catches rose from around 8,000 tonnes in 2001 to over 60,000 tonnes in 2004 (Ministry of Fisheries, Mozambique), an increase of almost eight times.

Mozambique's estimate of total marine capture production jumped from 24,000 tonnes (2002) to 78,000 tonnes (2003) (FAO, 2003c). The impact on food security statistics was a near doubling of per capita annual fish consumption statistics from 2.5 kg per person per year (average for 1999–2001) (FAO, 2003d) to 4.7 kg per person per year (average for 2003–2005) (FAO, 2006a).

## 2. Subsistence and local consumption

Fisheries and aquaculture provide fish either for direct consumption by fishers or producers and their families, or for sale at local markets. Small-scale fishers usually satisfy their subsistence needs first, before selling the rest of their catch. A study in the countries of the Southern African Development Community (SADC), found that coastal small-scale fishers usually first remove a portion of their catch for consumption, and then sell the excess. The amount that is retained is fairly constant, independent of the size or value of the catch (Gameiro and Wilson, 2003). Subsistence needs may be their first priority, but increasing catches enable them to gain benefits beyond these. Inland fisheries are especially important for food security as almost all production goes for human consumption rather than, say, animal feed (FAO, 1999). In Bangladesh and southeast Asia, small water bodies and reservoirs are important to the poor by providing a subsistence source of food, helping to reduce household expenditure (Sultana *et al.*, 2003).

## 3. Nutritional quality

Fish are a source of protein, micro-nutrients and essential fatty acids, providing an important complement to the predominantly carbohydrate-based diet of many poor people in developing countries. They are essential to a healthy diet and are particularly relevant to maternal health and child development.

Fish protein in particular, contains a good balance of essential amino acids which are found in similar proportions to those in human tissues, muscles and organs. These are able to complement proteins of vegetable origin, especially lysine-deficient proteins found in grains, thus enhancing overall Net Protein Utilisation.

Small, indigenous fish are particularly important for nutrition because they are eaten whole, 'bones and all', thereby providing a source of calcium and other micronutrients.[2]

### Box 2: Nutritional quality from small indigenous fish in Bangladesh

Fisheries are an important component of rural livelihoods in Bangladesh, where enhancement fisheries through stocking water ponds are common. Small indigenous fishes are a particularly important source of essential macro- and micro-nutrients, not available from other parts of diets of the rural poor, which are mainly based on rice. Because the small fish are often cooked and eaten whole, they provide an important source of calcium. Furthermore, some small indigenous fishes such as mola (*Amblypharyngodon mola*) and dhela (*Osteobrama cotio cotio*) contain more available vitamin A than any other freshwater fish.

Source: Wahab, 2003

## 4. Accessible food for the poor

Fish are often one of the cheapest protein sources available in developing countries. Small fish are especially important for poor consumers, as they can be purchased in small quantities at low cost. The same consumers often cannot afford to buy other protein sources which have to be bought in much larger quantities, such as chicken (Gameiro and Wilson, 2003). As an example, in Mozambique, fish is an important source of protein, in particular for the poorest segments of the population, providing 12 per cent of total protein consumed, compared to only 4 per cent for meat (INE, 2008).

Furthermore, fish are often processed locally and transported large distances, providing a low cost protein source for people who do not have access to fishing areas or fresh fish. Preserving is often carried out by smoking, drying or salting – methods which do not require complex or expensive technology.

While there are extensive trade networks for dried fish produced by small-scale operators throughout Africa and Asia, some industrial fisheries also provide low cost fish for urban dwellers. For example, horse mackerel caught off Namibia are widely available throughout southern Africa. During the Soviet era, Russian factory vessels plied the African coastlines catching and processing vast quantities of small pelagic fish such as sardinellas and mackerel both for local consumption and for its own markets in the USSR.

## 5. Income

38 million people worldwide are employed in fishing activity, 95 per cent of whom are in developing countries. The income from employment, and/or the financial revenues from the sale of fish, provide cash which can then be used for the purchase of other foodstuffs, or for other household expenditure.

In addition to the fishing activity itself, many associated enterprises also provide income and employment. 100 million people derive their income from fisheries and fisheries-related activities. For example, trading of fish, boat building and repair and processing (both traditional processing techniques and industrial/commercial processing oriented for export) are important. In many countries, women are traditionally involved in the processing and sale of fish. In Asia, aquaculture including the export-oriented processing industry, is an ever-increasing source of employment and income.

## 6. Reduced vulnerability

Fisheries can reduce vulnerability to food insecurity by providing a complementary source of food or income as part of a diversified livelihood strategy. Fisheries may provide food when other food sources such as agriculture are at a seasonal low, and *vice versa*. When there is little or no cost to the household, as with inter-tidal resources or species that do not require regular re-stocking in inland fisheries, their importance to the poor may be even greater.

Box 3 provides insights into the mixed farming/fishing subsistence systems in the Brazilian Amazon floodplains which are currently vulnerable to increasing fishing pressure from outside.

### Box 3: Mamirauá Sustainable Development Reserve, Amazonas, Brazil

The economy of the many settlements and small towns in the Amazon Basin is largely based on exploitation of the floodplain (*várzea*) fishery resources. The settlements in and around the *Mamirauá Sustainable Development Reserve* (MSDR) are typical. Local livelihoods depend on the maintenance of a dynamic balance between two mutually supporting systems: highly productive and biodiverse fisheries and low productivity subsistence agriculture. 80 per cent of revenues are generated from the sale of fish and fish products. However, the current delicate balance between fisheries and agriculture is vulnerable to the threat of commercial fishing operating from downstream centres of population growth (e.g. Manaus). Although fishing pressure in this remote region is still relatively low, a decline in productivity through increasing pressure on resources (especially the relatively few highly valued species) will irrevocably upset the balance and jeopardise the livelihoods of many thousands of várzea dwellers. Negative impacts have already been felt in other areas of the Amazon basin where traditional production systems have been displaced and this has led to considerable conflict.

Per capita consumption in the Amazon region is amongst the highest in the world. Average figures range from 48–50 kg per capita per year (various sources) for the várzea and riverine communities as a whole. Socio-economic surveys of MSDR communities showed higher consumption levels of 3–7.5 kg per household per day. These levels are much higher than Brazil's consumption as a whole (7.6 kg per capita per year), re-emphasising the regional importance – and vulnerability – of fisheries to food security and livelihood protection.

Source: Bostock (1998)

## 7. Fisheries and trade

Fish exports have become a major source of foreign exchange earnings for developing countries (Figure 1). Whilst concerns have been expressed over promoting fisheries exports from developing countries and the negative effect this could have on food security, exports can yield substantial benefits to domestic food security. They can do this in at least two main ways: at the macroeconomic level, the earnings from exports of high-value fishery products, such as lobster or shrimps, can be re-invested in imports of large quantities of less expensive food, which then becomes available to the wider population; and, at the microeconomic or household level, increasing fish exports can create jobs and raise incomes for poor people, who then have more to spend on food.

However, a new OECD study has examined the impacts of high-value fish exports on poverty and local food supplies in Africa (OECD, 2009). This focuses on concerns often expressed that the promotion of such exports to developed countries' markets may remove fish from African markets and consumers. Following an examination of macro-economic data from sub-Saharan countries, the study found that such concerns may be unfounded. However, the study also found that any positive impact on economic growth or poverty alleviation was equally elusive. The authors argue that this is due to the absence of any 'trickle-down' effect that might otherwise help distribute revenues from exports into welfare. The paper concludes by noting that trade revenues are dissipated before they have the chance to impact on any economic and/or human development indexes (OECD, 2009). This emphasises the need for macroeconomic policies to promote the redistribution of wealth and to focus on pro-poor growth from not just fisheries but all sectors of the economy.

## 8. Fisheries and economic growth

We have emphasised that fisheries provide food security through their direct contribution of food, livelihoods and incomes. Given an appropriate institutional context, fisheries can also contribute significantly to economic growth and wealth creation. Economic growth provides a pathway out of poverty for nations. Growth has accounted for about 80 per cent of the poverty reduction around the world since 1980, helping lift around half a billion people out of poverty and into food security.[3]

Unless, exceptionally, the fishery sector is very large relative to the rest of the economy, wide scale, sustainable poverty alleviation and food security are unlikely to be achieved through a single sector such as fisheries. Poverty and food security are macroeconomic governance issues to which the fisheries sector can only *contribute* as a creator of resource wealth (Cunningham *et al.*, in press). Nonetheless, lost wealth from fisheries, combined with the significant future risk of accelerating deterioration of fish stocks, are factors of particular importance to the developing world. Wealth foregone implies a significant lost opportunity to stimulate economic growth and provide pathways out of poverty.

Many Commonwealth states are fortunate in possessing extensive fish resources. While in some cases these already contribute significantly to their economies, in many others, especially in developing states, they are capable of contributing far more. For this to happen, the latent resource wealth (resource rent) needs to be turned into realised value-added. This can be achieved through royalties or taxation, or through the capitalisation of resource rents into the value of fishing rights. This extra value-added is of particular significance because the wealth on which it is based represents an investable surplus that can be used to generate economic growth throughout the economy. At the same time, it will also provide a basis to develop further wealth creation within the fisheries sector (by enhancing revenues, reducing costs or both). A welcome side-effect of this is that wealth generation will also lower exploitation rates and thereby increase the resilience of fish stocks to natural shocks such as climate change.

Governments need to better understand the macroeconomic contributions that the fishery sector is capable of making. This is a daunting challenge, particularly when starting from a position of overexploited fish resources and where the sector is often seen as making only marginal contributions to macroeconomic aggregates. Such a perspective is hardly surprising when, as is usually the case, its largest potential contribution to GDP – the resource rent – is dissipated.

In summary, fisheries provide many important benefits and contributions to food security, livelihoods and growth. However, fisheries are rapidly becoming overexploited and these benefits are being eroded. While the stagnation and decline in capture fisheries has been somewhat offset by aquaculture production, aquaculture systems are still a long way from satisfying demand. There remain many constraints to aquaculture development and to its enhanced contribution to food security, not least ensuring that its development is managed in socially and environmentally sound ways. Moreover, aquaculture development should not be at the detriment of capture fisheries, both inland and marine, that represent the vast bulk of existing and potential production and wealth for most of the world's poor. We will now take a closer look at the threats to sustainable fisheries and how we might begin to address them.

## Key Threats to Fish's Contributions to Food Security

The following *big numbers* speak for themselves about the substantial difficulties that global fisheries find themselves in today (Agnew *et al.*, 2009; MRAG, 2005; World Bank, 2006, Sumaila & Pauly, 2006):

- $1 billion – annual value of fish stolen from sub-Saharan Africa;
- $11–23 billion – annual value of global illegal unreported and unregulated fishing;
- $50 billion – cost to the global economy through fisheries mismanagement, i.e. failure to optimise efficiency through capturing resource rents;
- $30 billion – annual value of international cost reducing / capacity enhancing subsidies to fishing fleets.

In this section we explore some of the reasons behind such deplorable outcomes.

### 1. Overexploitation and poor management

Growing populations need more food for more people. Where fisheries remain open access, these are often exploited as social safety-nets, providing employment of the last resort. Although short-term benefits may well accrue in this regard (especially during periods of conflict when rural populations may migrate to the coast and come to rely on fisheries), rents are rapidly dissipated through overexploitation – too few fish being chased by too many fishermen. Eventually, the fishery will collapse, leaving the dependent population more vulnerable than they were before.

Many fisheries worldwide have followed similar evolutionary pathways,[4] moving from a situation of abundant resources to a period of expansion characterised by the use of more effective fishing technology and increasing catches, followed by stock decline and fishery closure – the classic picture of *boom* and *bust*. In these instances, which are experienced in both developed and developing countries, the management of fisheries often places a burden on taxpayers rather than making a positive contribution to national economies, including food security and poverty reduction.

## Box 4: Fisheries in Africa

The social and economic value of fisheries to Africa is vast, but remains largely unrecognised. For some 200 million people – about 30 per cent of the continent's population – fish is the main source of animal protein and micro-nutrition. Fisheries provide livelihoods to over 10 million men and women engaged in catching, processing and trade, many of whom are small-scale operators supplying food to local and sub-regional markets. Catching alone employs 2.6 million people, equivalent to 8 per cent of global fishers and more than the combined total of EU, the Americas, Australasia and the Pacific.

High-value fish exports from Africa have been increasing rapidly in recent years. In 2004 annual African exports amounted to US$4.3 billion, equivalent to 8 per cent of global value. Fisheries represent the leading agriculture export commodity for Africa (and indeed to developing countries globally), forming a significant element of some national economies. Yet, the sustainability of these benefits is under serious threat.

With a few notable exceptions, African fisheries governance is typically ineffective or absent, resulting in fisheries being over-exploited economically and often well beyond biologically sustainable limits. Policy objectives are often ill-directed and apparently sensible policy choices often have unintended, and sometimes perverse, consequences that policy makers may not appreciate. Amongst the more perverse of these is that fishers and policy makers regard fish resources as if they were free goods. Treating fisheries this way invites destructive rent-seeking behaviour. This factor has a bearing on policy choices and outcomes: for instance, where the resource is seen as 'free', it may be considered strategic to concede access rights to foreign fishers in order readily to access foreign exchange, even though in the absence of effective monitoring and enforcement, this is likely to result in fishing beyond environmentally sustainable levels and longer-term economic losses to the country.

Poor governance in general and ineffective fisheries management in particular, are putting this highly valuable food resource at severe risk. The underlying causes of this are institutional. There has been a wide-ranging failure to deal with free and open access to fisheries. More effective institutions are urgently needed to manage fisheries resources, generate wealth from them and direct this into profitable investment.

This susceptibility to over-exploitation is not unique to fisheries, but also occurs with other renewable natural resources. It is exacerbated in fisheries due to the lack of *effective use rights*. In fisheries, it is fuelled by the increasing global demand for fish products combined with weak governance, subsidies and a lack of effective resource management. Overexploitation represents a highly significant, yet little recognised, opportunity cost to the global economy. A study recently completed by the World Bank, presents empirical data on the level of natural resource wealth foregone due to the mismanagement of fisheries: globally this 'rent drain' amounts to some US$50 billion per year – equivalent to more than half the value of the current international seafood trade (World Bank and FAO, 2009) Specific studies in developing countries present a correspondingly bleak picture of fisheries in decline and rent dissipation typified by weak governance, open access and fleet overcapacity.

The facts are clear. The net contribution of fisheries to the global economy is currently negative. Aside from the gross loss of rent noted above, states provide a substantial volume of capacity-enhancing and cost-reducing transfers or subsidies, estimated at between *US$30 billion and US$55 billion per year* (respectively, Sumaila & Pauly, 2006; FAO, 1999). The important point is that with better management, fisheries have the potential to contribute far more to the global economy, developing countries (Box 4), local livelihoods and food security.

Relatively modest investment combined with political determination is needed to unlock the inherent resource wealth in fisheries and use this effectively to underpin national policy objectives on growth and poverty reduction.

## 2. Illegal Unreported and Unregulated (IUU) fishing

An estimated US$11–23 billion of fish is stolen as a result of illegal fishing worldwide (Agnew *et al.*, 2009). At least US$1 billion of this loss accrues to sub-Saharan Africa.

IUU fishing usually contributes to unsustainable impacts on both target species and the ecosystem. This is likely to reduce productivity, biodiversity and ecosystem resilience, which in turn can lead to a reduction in food security for artisanal fishers (see Figure 2). This is particularly important in those communities which are heavily dependent on fish as a source of animal protein, notably the coastal communities in countries such as Mauritania, Guinea Bissau, Guinea Conakry, Senegal, Liberia, Sierra Leone, Angola and other countries of West Africa and Somalia, Kenya, Tanzania and Mozambique in North Eastern, Eastern and Southern Africa. For example, in Liberia it has been reported that around 70 per cent of pre-conflict licensed catches were landed in Liberian ports. Illegal fishing rose to high levels with the breakdown in governance as a result of the conflict (estimated at 146 per cent of legal catches (MRAG, 2005)). Recent landings are significantly reduced, with important implications for the local availability of protein.

Agnew *et al.* (2009) demonstrated a significant inverse relationship on a global scale between the level of illegal and unreported fishing with World Bank indices of governance (government effectiveness, regulatory quality, rule of law and control of corruption).

**Figure 2: (a) Per capita consumption (kg/capita/year); (b) How this might increase through elimination of illegal fishing**

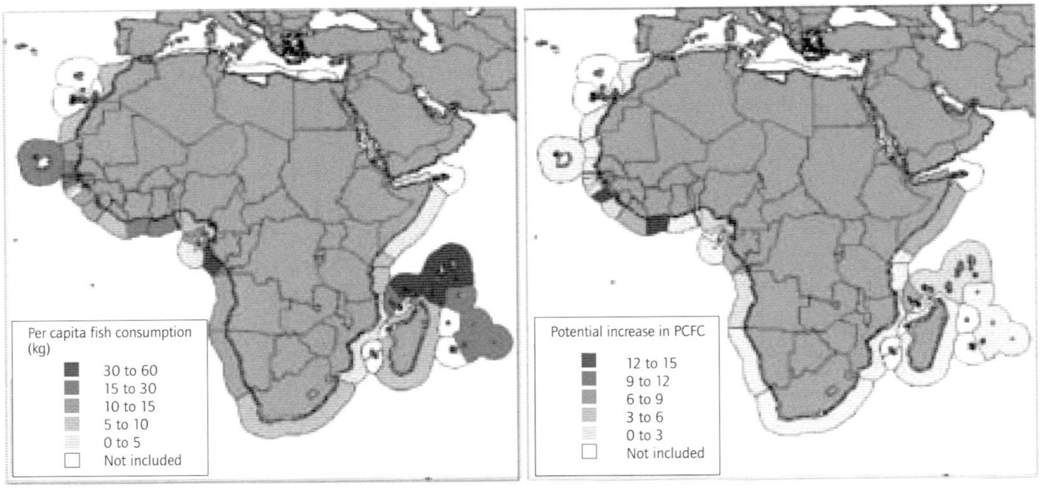


(a) (b)

Source: MRAG, 2005

Developing countries with poor governance are not to blame for illegal fishing, but they are more vulnerable to illegal activities – conducted both by their own fishers and by foreign vessels. In Africa, for instance, many coastal states license foreign vessels, and there is a significant reported illegal fishing problem from many of these. This represents a failure of control on behalf of the flag state as well as the coastal state.

## 3. Conflicts over resources

There are often many competing demands on finite fishery resources in both marine and inland fisheries. Conflicts arise as a result of competition for access, as well as cultural and political differences. Fisheries managers are therefore faced with issues of access and use rights. Where industrial fishing boats compete with local fishers for fishing areas or for the same fish stocks, the small-scale fishers usually lose out. This has serious implications for their food security. Protecting small-scale fishers' access rights through exclusion zones where the industrial fleet is banned from fishing, with appropriate enforcement backed by the Rule of Law, can help avoid such problems. For example, in northern Mozambique small-scale fishers and industrial boats had been in conflict over gear entanglement. A three mile exclusion zone was established to prioritise the small-scale fishers in this area. However, the zone is not always observed by the industrial vessels, emphasising the need for effective monitoring and enforcement.

In inland fisheries, competing demands for access and/or water use can lead to conflict, particularly where fish production may be a secondary use for water resources. In either case, the poor are often the most disadvantaged.

## 4. Habitat destruction and pollution

Habitat destruction and pollution threaten fisheries' productivity and are a particular problem for inland and inshore fisheries. Damage to or alteration of the habitat can reduce the production and wealth-generating potential of a fishery. For example, destruction of mangroves removes the nursery grounds of many fish species, and can result in reduced productivity of fish stocks.

The development of aquaculture also has implications for this. If not properly regulated and effectively controlled, aquaculture can result in habitat destruction from the establishment of production areas. For example, mangroves have commonly been cut down for the development of shrimp ponds; this can affect coastal protection and fisheries productivity. Aquaculture can also cause environmental pollution from effluents, such as organic pollution from feed and waste. The illegal use of antibiotics and hormones to control disease and increase growth rates can also result in them entering the surrounding environment if wastes are not properly treated. 'Escapees' from aquaculture production areas can have an impact also on local aquatic populations, especially if non-native species or genetically improved strains are cultivated, which may out-compete the native species in their natural environment.

## 5. Sustainability of inputs for aquaculture

Aquaculture rarely operates in isolation from capture fisheries systems. In particular, the source of juveniles or 'seed', and the source of feed for the farmed animals are important aspects of aquaculture whose impacts must be considered. The use of wild seed can have impacts on the wild populations and can affect productivity of wild stocks. Often wild seed is collected using very fine mesh nets, which may catch juveniles of other species, having far wider and destructive impacts on the aquatic ecosystem (Naylor *et al.*, 2000). This situation may be exacerbated by ineffective hatchery technology – particularly water purification – that leads to high mortality of hatchery seed and increasing dependency on wild stock.

The cultivation of carnivorous species (e.g. salmon) requires the presence of particular elements in the feed that are found in their natural diet. As a result, fish from wild capture fisheries are often used in the production of fish meal for use in aquaculture farms. Therefore aquaculture itself also has an impact on capture fisheries, with about 5–6 million tonnes of fish being used as direct feed in aquaculture worldwide (5.4 per cent of global capture production) (FAO, 2006b). We examine further aspects related to the efficiencies of protein conversion through aquaculture in section 6.4.

## Enhancing Fish's Contribution to Food Security

Many coastal states have fish resources that are large enough to contribute to food security at both local level and national, macro-economic level. At present, however, widespread overexploitation of these resources due to the absence of effective fisheries management policies means that the resources continue to deliver far less than their potential contribution to macroeconomic goals, local livelihoods and food security.

By far the biggest problem concerns the overexploitation of marine capture fisheries. Yet, international best practice, including some African experience (e.g. Mauritania, Namibia), is already providing compelling empirical evidence that well-managed fisheries can unlock fisheries wealth, producing a sustainable and renewable investable surplus which can be available to support economic growth elsewhere in the economy. As such, there are several ways through which the contribution of fish to food security can be maintained or increased. In this final section we examine some of these.

## 1. Sustainable fisheries through managed exploitation

If food security benefits derived from fisheries are to continue, fish stocks need to be managed and maintained at levels that ensure sustained productivity and wealth generation. This is far from a straightforward challenge, yet there are several cases where such 'best practice' speaks for itself, including from Africa (Box 5).

### Box 5: Potential benefits for Africa from better management

At the moment it is extremely difficult to estimate the potential payoff to Africa from good fisheries management. This is due mainly to the scarcity of information on the total value of African fish landings. However, it is clear from the little evidence available that the potential returns from good management are immense. Bioeconomic modelling undertaken recently in Mauritanian fisheries estimated that for the octopus fishery alone, effective management could produce a resource rent of some US$100 million per year (to be divided between the Government and the fishing industry). In the case of the Namibian hake fishery, estimates suggest that total annual rents generated by the rights-based management system are in the order of US$60–70 million per annum, of which between a quarter and a third is collected by the Government as royalties, with the remainder being left with the industry. It is notable that fisheries currently contribute 11 per cent to Namibian GDP compared with zero in the early 1990s. It is also suggested that scope exists to improve this performance.

These are clearly significant benefits, and are a result of the effective management strategies adopted and implemented by Mauritania and Namibia. They are also benefits that would be available in perpetuity if effective management remains in place. In Mauritania's case, rents at this level would cover 10 per cent of central Government expenditure from this one fishery alone; and Mauritania has significant other fish resources. Extending this kind of analysis to other African countries is clearly desirable so as to inform policy decisions but is currently impeded by the scarcity of information.

The importance of resource rent has long been acknowledged. By generating such rents through economically rational management, increases in value-added and sectoral contribution to GDP and growth can accrue. However, despite the successful adoption of such systems in some countries around the world, economics continues to have relatively little influence on fisheries policy. This lack of influence is particularly noticeable in developing countries – precisely where the contribution that effectively-managed fish resources might make to GDP is most urgently needed. Cunningham *et al.* (in press) conclude that focussing fisheries policy explicitly on the wealth of the resource is the only way to reverse the decline in fish resources and to ensure that the potential social and economic benefits from their exploitation are fully realised. Such a change in focus is especially urgent in developing countries, where the need to generate maximum sustainable benefits from scarce fish resources is the greatest.

The key requirement to increase the adoption of such economically rational fisheries management is to convince policy-makers to focus explicitly on the wealth-generating potential of fish resources. Such a *wealth-based* focus provides a general policy framework within which other approaches based on rights, incentives and ecosystems, may be nested. This approach is likely to prove more effective in influencing policy, especially in situations where rights-based systems either will not work or are politically unacceptable.

It is nonetheless important to emphasise the key role of well-defined use rights in reforming fisheries management. Rights-based management can help achieve sustainable catches and raise profitability (Leal, 2005). Indeed, as with many other renewable natural resource systems, effective management solutions embrace well-defined, enforceable and transferable use rights in the resource itself. Rights can help secure the traditional access rights of small-scale fishers as well as promoting efficient use of available fishing opportunities by larger-scale fleets. Fisheries use rights are complex and multidimensional (Deacon, 2009) – to be effective and efficient in capturing and distributing wealth, they must be seated within an enabling institutional and political environment.

Despite a growing body of compelling empirical evidence of their value in creating fisheries wealth, there are political economy constraints to rights and their allocation, which some commentators equate to 'handing over public assets to the private sector' or 'fencing off the commons'. It can take many years (i.e. beyond normal political horizons) to eliminate errors of the past and for rights (when properly instigated) to begin to bear perpetual fruit. Furthermore, as alluded to above, rights may only be effective when coupled with wide-ranging institutional reform which in most cases extends well beyond traditional sectoral boundaries (e.g. effective banking, legal and fiscal institutions).

Nonetheless, it is now without doubt that rights-based management offers a solution for long-term, cost-effective management of renewable natural resource systems.

## 2. Institutional reform

It has already been noted that the extent to which fisheries are able to operate on an efficient, equitable and sustainable basis is directly linked to existing fisheries governance and management institutions. The World Bank in its seminal work on wealth of nations (World Bank, 2006) demonstrated that the transformation of natural capital (such as fisheries and forest assets) into other forms of capital (produced, human and governance) depends on a set of institutions capable of managing these natural resources, generating resource rents, and directing these rents into profitable investments including welfare. In situations where fisheries have operated successfully, such institutions have been instrumental in unlocking the potential value of fisheries. Resource policy, fiscal policy and the political economy all have a major role to play in this transformation.

At the national level, institutional reform is likely to involve politically complex decisions involving tackling allocation and capacity reduction. Although the tools that can support institutional change are available (e.g. rights, legal and fiscal reform), some of these may require an about-turn in political thinking. Reform can represent a radical departure from 'business-as-usual' approaches to conventional fisheries management. To be successfully applied, political courage is needed as well as commitment by the fishing industry.

At the international level, the incoherencies and often contradictory nature of policies is a hindrance to international fisheries development. OECD (2006) recognises that:

> *Policy coherence both within and between OECD and non-OECD countries has a major impact on the livelihoods and poverty status, economic performance, social conditions and food supply of millions of people throughout the world; policy incoherence occurs in all policy domains in fisheries (environment, technology, economic, social and governance).*

## 3. Management of conflicts over resource access

National development and fisheries policies should recognise the important contribution of all fisheries, but particularly the small-scale sector, to food security. Secure access to sustainable resources must be maintained. Policy makers need to be more aware of the different contributions that fisheries make to food security and ensure management, access rights and governance structures help maintain these benefits. It is noteworthy in this context that the Code of Conduct for Responsible Fisheries (FAO, 1995) requires that states protect the rights of small-scale fishers, as well as their preferential access to inshore fisheries through policies and laws. This is specifically recognised in Article 6.18 of the Code, which states that:

> *Recognizing the important contributions of artisanal and small-scale fisheries to employment, income and food security, States should appropriately protect the rights of fishers and fish workers, particularly those engaged in subsistence, small-scale and artisanal fisheries, to a secure and just livelihood, as well as preferential access, where appropriate, to traditional fishing grounds and resources in the waters under their national jurisdiction.*

Evidence from the vast majority of developing country fisheries (with some notable exceptions) seems to show shortcomings in the way this important aspect of the Code has been interpreted. Rather than 'appropriately protecting preferential access rights' with a view of generating sustainable benefits from fisheries, approaches to access have tended to go the other way, favouring an unsustainable (economically, socially and environmentally) safety net function. Even FAO (FAO, 2005b, p. xiv) has argued that:

> *in the context of generally poor communities, it is impossible to exclude people living on the edge of survival from fishing without creating alternative sources of food and livelihoods. Exhortations about reducing pressure on fisheries resources are futile as hungry people will choose, quite reasonably, to survive in the short-run rather than to preserve or rebuild a resource that they might not otherwise survive to benefit from.*

The ability of fisheries to operate on a sustainable basis, producing social and economic benefits, is directly linked to the fisheries management arrangements that govern fisheries. In the few instances (globally) where fisheries have operated sustainably and contributed to economic development rather than detracting from it, unlocking the real economic value of fisheries through institutional reform has been a central feature of their success.

## 4. Markets and trade

Effectively-functioning markets are fundamental to ensuring global food security. Markets need to continue to be liberalised through the Doha Development Round of trade negotiations in order to provide better market access opportunities for developing countries. This can help increase the contribution that fisheries can make to developing countries' economies through exports, although this must be managed in a sustainable and equitable way and ensure that resources are exploited within their sustainable limits.

In particular, developing countries should be supported to help them meet the standards and criteria set by importing countries and retailers – for both mandatory and voluntary requirements. However, any increase in fish exports from developing countries should be monitored and increased prices on local markets or reductions in local availability of fish should be avoided.

Market-based instruments such as certification systems and ecolabels that are capable of communicating environmental and/or social benefits to consumers provide a means of harnessing the power of markets towards promoting sustainability of both fisheries and aquaculture systems. Increasingly, governments and industry are recognising their importance, and there are an increasing number of initiatives in the industry and retail sectors. Such schemes are likely to have a positive impact on food security by ensuring the effective management of resources, the environment and the impact on local communities.

## 5. Enhancement and aquaculture

Stocking water bodies with juveniles can increase production and availability of mature fish, generating benefits for the poor and increasing food security. As a form of small-scale aquaculture, this can be particularly effective in inland systems. For example, in Laos, stocking increased the production potential of small reservoirs and provided a source of food for villagers in times of need (Garaway *et al.*, 1997).

The development and growth of aquaculture also has the potential to increase the contribution of fish to food security, both through increased production, as well as the indirect benefits that are generated such as employment on aquaculture farms and in processing plants.

Aquaculture offers notable efficiencies when compared with traditional agriculture systems: it takes roughly four times the amount of feed to produce one kilo of beef as it does one kilo of fish. This is because some farmed fish are particularly efficient at producing protein – carp need less than 2 kg of feed grain to produce 1 kg of meat, whereas grain-fed cattle require around 8 kg and pigs need 3–4 kg. Under such circumstances, there is the potential for future protein needs of consumers to be increasingly met from fish. China accounts for one third of global fish production, with two-thirds of that from aquaculture. China has developed highly efficient fish farming systems involving carp polyculture integrated with agriculture which offer significant potential for application elsewhere.

Despite this, there are a number of issues that must be taken into account and addressed for aquaculture to contribute to food security in a socially and environmentally sustainable manner. The environmental impacts of aquaculture must be addressed and minimised – including the impacts of habitat destruction, effluents and pollution, escapees from aquaculture farms, and impacts on local capture fisheries of wild-caught seed.

Although carp feed on vegetable and agriculture products, other popular aquaculture species are carnivorous and require large inputs of wild fish in their feed. For example, salmon consume over 8 kg of wild fish for every 1kg of salmon produced (Pinto & Furci, 2006). There is therefore a need for the development of other feed substitutes so that aquaculture can reduce its dependence on wild capture fisheries production in its feed and be sustainable in the long term.

Aquaculture also needs to reduce its dependence on wild systems as a source of seed and juveniles for cultivation. Although a number of advances have been made in closing the reproductive cycle and the ability to breed species entirely in captivity, there is still a dependence on wild seed. This has impacts on the productivity of wild stocks and the surrounding ecosystems and more research is needed to reduce the dependency of aquaculture on wild seed.

Aquaculture therefore represents a significant growth opportunity. However, it needs to be considered alongside, not as a replacement for, capture fisheries, which are vastly more important than aquaculture in global economic terms, and its impacts must be addressed to ensure it is sustainable in the long term.

## 6. Integrated initiatives in developing countries

Often, solutions to the problems in fisheries lie outside the fisheries sector. Integrated, cross-cutting interventions are necessary to address such issues. For example, infrastructure problems (lack of electricity, poor roads, risky business environment, landing centres and so forth) are often cited as one of the key barriers to effective fisheries and associated trade.

Such weaknesses need to be addressed, but must be done so within the context of an effective management system.

### 7. Gender

The role of women (and in some cases children) in fisheries, particularly in post-harvest processing for traditional and export markets is enormous. The different and varied roles of both men and women therefore need to be considered in interventions aiming to address fisheries issues.

## Conclusions

The recent increases in global food prices have focused the attention of governments around the world on short-term supply issues and long-term challenges to our food system. Rising global demand, climate change, high oil prices and new pressures on land such as biofuels are undermining global food security. These pressures are compounded by trade-distorting subsidies and protectionist policies imposed by the US, EU and other countries. In this Chapter we have shown that fisheries and aquaculture contribute to food security in two principal ways. First they provide a high quality food containing a range of nutrients essential to a healthy diet and particularly relevant to maternal and child development. Second, they generate wealth that provides livelihoods and that can fuel sustained economic growth. Being a renewable natural resource, benefits from fisheries may be generated indefinitely and in many cases enhanced. Achieving this ultimate goal is complex. The challenge is to put in place policies and structures to enable more effective fisheries management systems. Better management of wild-caught fish (marine and freshwater) and careful management of the growth of aquaculture activities offer massive potential to contribute sustainably to food security, growth and trade to the benefit of all Commonwealth economies.

# References

Agnew, D.J., Pearce, J., Pramod, G., Peatman, T., Watson, R. *et al.* (2009) Estimating the worldwide extent of illegal fishing. *PLoS ONE* 4(2): e4570 doi: 10.137/journal.pone.0004570.

Bostock, T. (1998) *Mamirauá Sustainable Development Reserve, Amazônas, Brazil – Fish Processing and Marketing*. Report for DFID and MSDR.

Chuenpagdee, R.L., Ligouri, L., Palomares, M.D. and Pauly, D. (2006) *Bottom-up, global estimates of small-scale marine fisheries catches*. Fisheries Centre Research Reports 14(8). 112 pp.

Cunningham, S., Neiland, A.E., Arbuckle, M. and Bostock, T. (in press) Wealth-based fisheries management: using fisheries wealth to orchestrate sound fisheries policy in practice. *Marine Resource Economics*, in press.

Deacon, R.T. (2009) *Political Economy of Natural Resource Reform: Paper presented at Workshop on the Political Economy of Fisheries Reform*, Property and Environment Research Centre (PERC), Montana, May 2009.

DFID & MRAG (2008) *Fisheries and International Trade Policy Brief 7*. 4 pages, London: MRAG. Available at *http://www.mrag.co.uk/Documents/PolicyBrief7_International_Trade.pdf*

FAO (1995) *Code of Conduct for Responsible Fisheries*. Food and Agriculture Organization of the United Nations (Rome: FAO).

FAO (1999) *Future Challenges in World Fisheries and Aquaculture*. 23rd Session of the Committee on Fisheries. 15–19 February 1999 (Rome: FAO).

FAO (2002) *The State of Food Insecurity in the World 2001* (Rome: FAO).

FAO (2003a) *Assessment of the World Food Security Situation*. 29th Session of the Committee on World Food Security, 12–16 May 2003 (Rome: FAO).

FAO 2003b. *Report of the twenty-fifth session of the Committee on Fisheries. Rome, 24–28 February 2003*. FAO Fisheries Report. No. 702 (Rome, FAO).

FAO (2003c) *Fishery statistics – Capture production*. FAO Yearbook Vol. 96/1. (Rome: FAO).

FAO (2003d) *Fishery statistics – Commodities*. FAO Yearbook Appendix I Fish and Fishery Products – Apparent Consumption.

FAO (2005a) *FAO/Worldfish Center Workshop on Interdisciplinary Approaches to the Assessment of Small-Scale Fisheries*. 20–22 September 2005. FAO Fisheries Report No.787. Rome: FAO.

FAO (2005b)). *Increasing the contribution of small-scale fisheries to poverty alleviation and food security*. FAO Technical Guidelines for Responsible Fisheries. No.10. Rome: FAO

FAO (2006a) *Fishery statistics – Commodities*. FAO Yearbook Appendix I Fish and Fishery Products – Apparent Consumption.

FAO (2006b) *Use of Fishery Resources as Feed Inputs for Aquaculture Development: trends and policy implications*, by A.G.J. Tacon, M.R. Hasan and R.P. Subasinghe. FAO Fisheries Circular No. 1018. Rome: FAO.

FAO (2009) *The State of World Fisheries and Aquaculture 2008* (Rome: FAO).

Gameiro M.F.S. & Wilson, J.D.K. (2003) *The importance of marine fisheries to coastal community livelihoods in SADC countries* (Maputo: Kusi Consultores Ltd).

Garaway, C., Lorenzen, K. and Chamsingh, B. (1997) *Reservoir Fisheries Management in Savannakhet Province, Lao PDR* (London: MRAG).

INE (2008) *Food Insecurity Assessment Based on Food Consumption Statistics derived from the 2002/03 Mozambique household budget survey*. Summary Report, March 2008. Funded by the European Union through the EC-FAO Food Security Information for Action Programme (Maputo: Instituto Nacional de Estatística).

Leal, Donald R. (2005) *Evolving Property Rights in Marine Fisheries* (Lanham: Rowman & Littlefield).

MRAG (2005) *Review of Impacts of Illegal, Unreported and Unregulated Fishing on Developing Countries*. Final Report for DFID (London: MRAG).

Naylor, R. L. Rebecca J. Goldburg, Jurgenne H. Primavera, Nils Kautsky, Malcolm C.M. Beveridge, Jason Clay, Carl Folke, Jane Lubchenco, Harold Mooney & Max Troell (2000) Effect of aquaculture on world fish supplies. *Nature* 405, pp. 1017–1024.

OECD (2006) *Fishing for coherence: fisheries and development policies* (OECD Publishing).

Pinto, F. & Furci, G. (2006) *Salmon Pirana Style: Feed conversion efficiency in the Chilean salmon farming industry*. Study by the Terram Foundation. English language summary, original study in Spanish available at *www.terram.cl* (Santiago de Chile: Terram).

Sultana, P., Thompson, P. and Ahmed, M. (2003) *Understanding Livelihoods Dependent on Inland Fisheries in Bangladesh and Southeast Asia. Final Technical Report* (DFID/FMSP Project R8118). London: MRAG.

Sumaila, R. and Pauly, D. (2006) *Catching More Bait: A bottom-up re-estimation of global fisheries subsidies (2nd version)*. Fisheries Centre Research Reports, Vol 14 No 6 (University of British Colombia).

Wahab, M.A. (2003) Small indigenous fish species of Bangladesh: Potentials for culture and conservation. pp. 1–12. In: Wahab, M.A., Thilsted, S.H. and Hoq, M.E. (eds.), *Small Indigenous Species of Fish in Bangladesh*. Technical Proc. of BAU-ENRECA/DANIDA Workshop on Potentials of Small Indigenous Species of Fish (SIS) in *Aquaculture & Rice-field Stocking for Improved Food & Nutrition Security in Bangladesh*. 30–31 October 2002 (BAU, Mymensingh, Bangladesh. Bangladesh Agricultural University, Mymensingh 2202, Bangladesh).

Walmsley, S., Evison, S. & Cansano, J. (2002) *Mozambique Artisanal Fisheries Information Systems. Overview Report Volumes 1–3. Volume 1: Summary of institutions and information systems*. Southern African Development Community: Regional Fisheries Information Systems Project.

World Bank (2006) *Where is the Wealth of Nations?* Washington: World Bank.

World Bank and FAO (2009) *The Sunken Billions: The Economic Justification for Fisheries Reform* (Washington: World Bank and FAO).

## Notes

1. 1996 World Food Summit.
2. Evidence from a project commissioned jointly by the FMSP and Aquaculture and Fish Genetics Research Programme (AFGRP).
3. Speech: 'Growth at the heart of development' – Speech by Douglas Alexander, Secretary of State for International Development, on Monday 31st March 2008, at the Institute of Directors, London.
4. Expert Consultation on Low-cost Fisheries Management Strategies and Cost Recovery, Georgetown, Guyana, 4–7 September 2007.

# 10 Fisheries Supply Chain Issues for Developing Countries

Charlotte Tindall

## Introduction

Fisheries products are some of the most heavily traded commodities in the world. Developing countries already contribute significantly to global transfers, but the potential for export in wild-caught fisheries has not yet been fully realised.

The huge demand for fisheries products has consequences for the sustainability of fisheries worldwide. The international market can play an important role in redressing overexploitation by creating a demand for fish from certified sustainable and legal sources. This is being achieved through a range of supply-chain mechanisms. However, developing countries face constraints in reaching existing and new market standards, and the benefits accruing to them may be reduced as a result.

While fisheries in developing countries provide a rich resource for many, the sector is not immune from poverty. Trade has the potential to support poverty alleviation if supported under the right conditions.

This chapter describes the state of play in fisheries supply chains, particularly the mechanisms that have been used to promote sustainability and legality. It asks what effects these are having on developing countries and whether they can be used to promote poverty reduction.

## Developing Country Engagement in Fisheries Trade: Benefits, Costs and Lost Opportunities

Global trade in fish was worth $86 billion in 2006 and approximately half of this trade originated from developing countries (FAO, 2008). In fact trade in fish from developing countries outstrips other commodities, including coffee and rubber. The map illustrates some of the key fish exporting developing countries and those that are also Commonwealth countries.

International trade in fish offers a number of benefits to developing countries, including foreign exchange, employment and in some cases higher prices for producers. Net revenues from fish products for developing countries were US$25 billion in 2007 (FAO, 2008). However, in the wild-caught sector a large proportion of production (especially that from small-scale fisheries) cannot reach export markets due to the lack of capacity to achieve necessary hygiene, sanitation and quality standards.

Where export potential is realised, there can be negative as well as positive implications of trade. The cost-benefit balance will depend on the local context and policies. In Uganda, demand from international markets and trade liberalisation has led to significant exports of Nile Perch from Lake Victoria. This has provided foreign exchange revenue and livelihoods for domestic fishers, traders and processors. However, high prices have attracted many people into the fishery and with insufficient controls this has led to overexploitation. Some fishing communities have also complained that due to a lack of investment in infrastructure at landing sites they have been unable to benefit fully from export (Keizire, 2004). Uganda also exports mukene (a small, low value sardine-like fish) from Lake Victoria for the use of fishmeal. While it is thought that here in Uganda this has threatened food security for the poor, fisheries export from other countries (including India and Mauritania) has had no impact on *per capita* fish consumption as either the species exported are not consumed locally or aquaculture and fish imports provide sufficient availability on domestic markets (FAO, 2003).

**Figure 1: Important developing country fish exporters (both capture and aquaculture)**

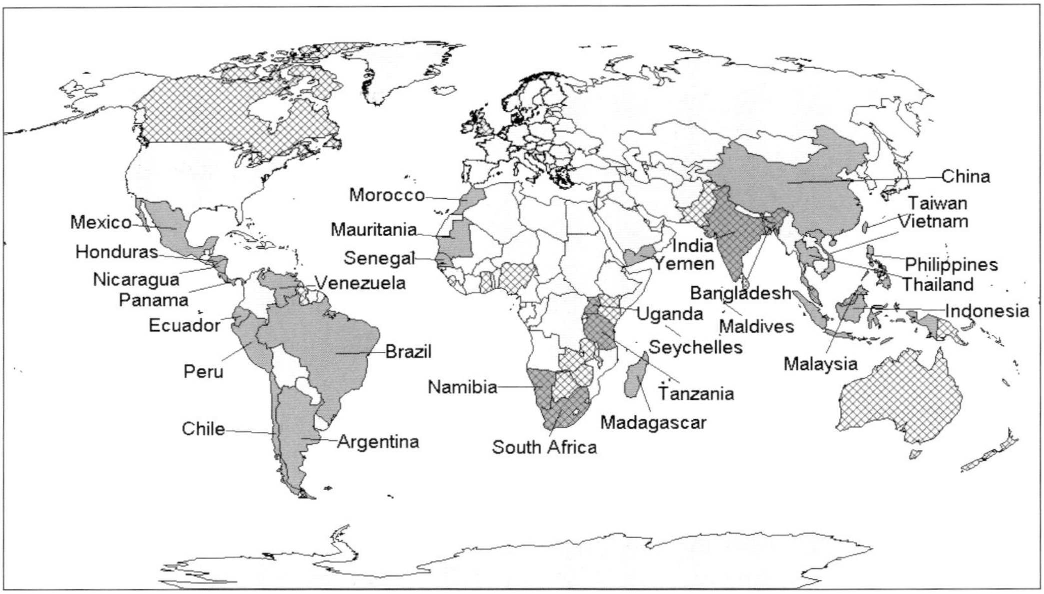


Commonwealth countries shown in cross-hatch. Source: MRAG (2008b)

## Sustainable and Secure Supplies: Implications for Developing Countries

One of the consequences of the huge demand for fish products on the international market is its contribution to overexploitation. The latest FAO status report suggests that 75 per cent of fisheries are fully or over exploited (FAO, 2007b). There are particular concerns for developing country fisheries where governance is often weak and the capacity for fisheries management low. Illegal, unreported and unregulated fishing contributes to the overexploitation of resources and, for example, robs sub-Saharan countries of an estimated $0.9 billion per year (MRAG, 2005; FAO, 2002).

The international market can play an important role in addressing overexploitation of resources by creating a demand for fish from sustainable and legal sources. This is being achieved through a range of mechanisms including NGO awareness initiatives, private sector buying policies, fisheries certification and traceability schemes. These initiatives have further implications for developing countries wishing to export fish products, as discussed below.

## NGO awareness initiatives

While FAO has warned of overexploitation of fisheries stocks for a number of years, it has been the advocacy efforts of environmental non-governmental organisations (NGOs) such as WWF and Greenpeace that have brought the issue to the attention of consumers. In recent years, such organisations have also raised awareness of the astounding amount of illegal fish that is traded and of the consequences for the marine environment.

Many of these NGOs provide advice to consumers on the sustainability of the trade in individual fish species. For instance WWF[1] and the Marine Conservation Society (MCS) provide guides on 'fish to eat' and 'fish to avoid'. Other organisations that provide such lists include the Monterey Bay Aquarium and the Sustainable Fisheries Partnership. Greenpeace publishes an International Seafood Red list and IUCN provides a 'red-list' of endangered fish species. Table 1 compares the criteria used by some of these schemes.

**Table 1: Comparison of NGO 'red-list' schemes**

| **Criteria** | **NGO advice** | | | | | |
|---|---|---|---|---|---|---|
| | Marine Conservation Society (MCS) | Fish source | Greenpeace International Seafood Red List | Monterey Bay Aquarium | IUCN Endangered Red List | WWF guides (SCA methodology as of August 2007) |
| Stock status/abundance | • | • | • | • | • | • |
| Geographic range | | • | | | • | |
| Vulnerability of species to fishing | • | | • | | | • |
| Environmental impacts | • | • | | • | | • |
| Fishing method | • | | • | | | • |
| Management measures | • | • | | • | | • |
| Use of scientific advice in management | | • | | | | |
| Legality | | • | | | | |

Species originating in developing country fisheries fall on these lists both as 'fish to eat' and 'fish to avoid'. The Marine Conservation Society guide suggests avoiding golden eye perch, snapper and grouper (from tropical coral reef fisheries) and marlin, swordfish and albacore tuna from the Indian Ocean. However, they also encourage consumption of fish from certified fisheries such as the MSC South African Hake fishery or where small-scale fisheries use low impact gear, for example pole and line caught tuna from the Maldives or the Central and Western Pacific, and line caught dolphin fish (mahi mahi) originating from Ecuador and Peru.[2]

## Private sector procurement policies

Buyers of seafood have had to respond to concerns expressed by the media, NGOs, governments and consumers by adopting voluntary standards or buying policies. Box 1 illustrates some of the commitments that have recently been made by retailers in Europe and the USA. Retailers often include other aspects in their procurement policies such as specification of certain gear types, exclusion of NGO 'red-listed species' and traceability.

### Box 1: Private sector commitments towards sourcing sustainable fisheries products

- **Sainsbury's** (UK retailer) has made a commitment to source all its wild-caught fish from sustainable sources by 2010.
- **Metro** (German retailer) had a target to stock 80 MSC products by 2008. There are currently 32 MSC-certified products on the shelves.
- In the Netherlands, the Dutch retail consortium **Centraal Bureau Levensmiddelenhandel** (made up of 25 chains of retailers) has set targets to sell only MSC-certified (or Global-GAP certified) seafood from 2011.
- **Marks & Spencer** (UK retailer) has committed to sourcing 100 per cent of their fish from sustainable sources (MSC certified or equivalent) by 2012.
- **Wal-Mart** (US retailer) has announced that they intend to shift their supplies of wild caught fresh and frozen fish to MSC certified fisheries by 2009–2011. Their UK counterpart Asda has matched this commitment to source MSC certified fish by 2010.

In order to sell into retailers within international markets, developing countries need to be able to fit these criteria. Supermarkets will often have preferred suppliers that are able to reach these standards and provide the relevant documentation. This may exclude smaller companies or producers that are not linked with larger suppliers, with inevitable impacts on artisanal fishing communities.

## Certification schemes

Many of the retailers' commitments make reference to the Marine Stewardship Council (MSC) certification scheme, which was one of the first to emerge (MSC, 2002). The scheme uses an eco-label to provide a sustainable choice for consumers and it is now one of the most recognised schemes, used in markets throughout Europe, America and now in the Far East (e.g. Japan). However, there are other certification schemes in operation. Box 2 provides a brief summary of these, and Table 2 compares their different elements.

### Box 2: Key fisheries certification schemes

**Marine Stewardship Council (MSC):** *www.msc.org*
MSC was established in 1997 and has emerged as the leading fisheries certification scheme, covering 8 per cent of global wild-caught fish production. It has certified 39 fisheries, is assessing 101 more, and had 2000 MSC-labelled products as of February 2009. It is considered a robust assessment scheme which assesses stock status, impact of fishing on the wider ecosystem and effective management systems. MSC are currently piloting an assessment approach for data-deficient fisheries, which may overcome some of the barriers experienced by developing country (and small-scale) fisheries.

**Friend of the Sea:** *www.friendofthesea.org*
Friend of the Sea is a rapid assessment scheme which uses published data on stocks to determine whether they are over-exploited or whether they can be certified. The assessment does not use additional data or include any criteria on fisheries management. Since the assessment process is more rapid than for the MSC it is more affordable and attractive to developing countries. However, it is not considered robust or credible by some actors, e.g. WWF. The scheme has gained market access with retailers in Italy and Spain.

**Fair Fish:** *www.fair-fish.ch*
The Fair Fish approach has been piloted within a multi-species artisanal fishery in Senegal. The assessment includes environmental, social and animal welfare criteria. Although trading relations were initially set up with a Swiss supermarket with minimum prices and a price premium arranged (similar to the Fairtrade model) the agreement with the supermarket fell through possibly due to low volumes or a lack of demand from the market.

**Naturland:** *http://www.naturland.de/naturland_fish.html*
Naturland is predominantly an organic label with customer recognition in Germany and Switzerland. It has recently developed a standard for sustainable wild-caught fisheries which includes ecological, social and economic criteria. It is currently testing the approach on a Nile Perch fishery in Lake Victoria, Tanzania (see Box 4) (Naturland, 2008).

**Seafish Responsible Fishing Scheme (RFS):** *http://rfs.seafish.org/*
Seafish RFS certifies vessels that comply with good fishing practices (capture, storage, handling and traceability), vessel criteria, crew competence and reduced impact on the environment (e.g. reducing marine litter and discharges). The scheme originated in the UK, but is currently being tested on tuna vessels operating in Sri Lanka.

**Fish4Ever:** *www.fish-4-ever.com/*
Fish 4 Ever sources fish from 'small-scale' fishing operations with low environmental impacts targeting tuna, sardines and anchovies off the coast of Spain and Portugal. It has recently certified skip-jack tuna from the Maldives and has developed a 'Fair Fish' logo to represent payment of a 10 per cent premium to tuna fishermen.

**KRAV:** *http://www.krav.se/*
KRAV certification covers a range of organic products, but has recently introduced a wild fisheries standard. This has been applied to cod fillet, tinned herring, fresh shrimp and fresh crabs from Scandinavian waters, but has not yet been applied to developing countries.

These certification schemes vary in the way in which they assess fisheries and what criteria they include. For instance ecological sustainability of the fisheries can be assessed according to the vessel practices, the state of the fisheries stock and the effectiveness of fisheries management. MSC is currently the only scheme which assesses fisheries management as part of its ecological criteria, and others such as the Seafish Responsible Fishing Scheme (RFS) and Fish4Ever look at vessel practices while Naturland and Friend of the Sea focus their considerations on the current state of the stock.

Social and economic criteria are included in some of the schemes. MSC includes some social elements but these are not core to the standard and form a marginal part of the performance indicators used to measure the criteria. Other standards address social issues in more detail. For instance Naturland include labour issues (e.g. working hours, minimum pay etc) and wider socio-economic welfare of workers (e.g. access to health and education). The Fairfish pilot in Senegal included a price premium to invest in community development and included criteria to promote the role of community organisations, and Fish4Ever has also introduced a 10 per cent premium for Maldives tuna fishermen. Fairtrade has recently expressed interest in the fisheries sector, particularly focusing on small-scale fisheries but also addressing labour issues within industrial fisheries (Box 3).

The schemes also differ in the markets that they target. While MSC is the most widely accepted in Europe and the US, it is not always popular among producing countries. In some countries, such as India and Sri Lanka, the highly rigorous certification schemes are often seen as technical barriers to trade. For this reason UK buyers of tuna from Sri Lanka have been exploring the use of the Seafish RFS as an alternative or stepping stone to a more inclusive certification. Naturland has a significant market presence for organic food in

Germany and Switzerland, while Friend of the Sea is gaining market acceptance in some of the Southern European countries (e.g. Italy and Spain).

### Box 3: Fairtrade approach to fisheries

Fairtrade has recently been looking at the opportunities to apply its model to fisheries. Although the approach is most suitable to small-scale fisheries, it could also be applied to industrial fisheries to help improve labour standards. One of the options open to Fairtrade is to team up with a fisheries certification scheme such as the Marine Stewardship Council (MSC) so that the environmental sustainability of the fishery is assured. Fairtrade would be able to build on this, ensuring fair trading practices. The Fairtrade model sets a minimum price for producers, which helps them deal with fluctuating markets, and it establishes a price premium for local cooperatives that can be used to promote local development projects. The constraints for Fairtrade are the limited number of developing country fisheries certified by MSC, and the lack of export quality products from small-scale fisheries.

Source: MRAG, 2008b

FAO has provided guidelines on eco-labels, setting out minimum criteria and procedural considerations (FAO, 2005a). These focus only on ecological criteria, although complementary guidelines on responsible fish trade are currently being discussed (FAO, 2007a). DFID, GTZ, UNEP and the OECD are among the donors and international agencies that are concerned about developing countries' engagement in international fish trade and their ability to reach export standards.

## Application of certification schemes in developing countries

Reaching supply chain standards, such as certification, provides opportunities for developing country fisheries to enter international markets, and add value to their products. For example tuna from the Maldives is actively marketed by a number of UK retailers as coming from 'small-scale artisanal fisheries'. However, since the introduction of certification schemes, the bias has been in favour of developed country fisheries. Out of the 39 fisheries certified by MSC, only three are from developing countries. Its application to developing countries has been restricted by the significant scientific data required to undertake the assessment to prove sustainability, and the cost of assessment.

However, in recent years there has been more effort to promote the certification of developing country fisheries. MSC has developed a new risk-assessment methodology for data deficient fisheries (GASSDD - Guidance for the Assessment of Small-Scale and Data-Deficient Fisheries) which is being piloted in six fisheries (five in developing countries, see Table 2). The methodology uses local and expert knowledge to assess the scale, intensity and consequence (i.e. impact) of the fishery on the ecosystem, and assess characteristics of

### Table 2: Key existing certification schemes for capture fisheries

| Certification scheme | Main markets | Sustainability criteria: Ecological – Vessel practices | Ecological – Fisheries stock | Ecological – Fisheries management | Social | Economic/Trade | Application to developing countries |
|---|---|---|---|---|---|---|---|
| **MSC** | Europe (UK, Germany, Switzerland, France), US Japan, Australia, South Africa | • | • | • | | | Mexico – rock lobster<br>South Africa – hake<br>Vietnam – ben tre clam<br>*Pilots*<br>Mauritania – mullet<br>Gambia – sole<br>Argentina – mullet<br>Ecuador/Peru – dolphin fish (mahi-mahi)<br>India – Indian oil sardine |
| **Friend of the Sea** | Spain, Italy | • | • | | | | Senegal – tuna<br>Sri Lanka – tuna<br>Morocco – sardines and mackerel<br>Vietnam – squid and cuttlefish<br>Philippines – tuna<br>Maldives – tuna<br>Indonesia – shrimp<br>Costa Rica – shrimp<br>Madagascar – shrimp |
| **Naturland** | Germany, Switzerland | | • | | • | • | *Pilot*<br>Tanzania, Lake Victoria – Nile perch |
| **Fish4Ever** | Europe, US | • | | | | | Maldives –tuna |
| **Seafish RFS** | UK | • | | | | | *Pilot*<br>Sri Lanka – tuna |
| **Fair Fish** | Switzerland (but not currently marketed) | • | • | | • | • | *Pilot*<br>Senegal – multi-species artisanal fishery (co-certified by Friends of the Sea) |
| **Krav** | Sweden | • | • | | | | *Only applied in Sweden to date* |

the fish species in determining the productivity of the stock and its susceptibility to fishing pressure.

Naturland's pilot for the Nile Perch on Lake Victoria also provides a potential well suited to the developing country context and puts equal emphasis on both social sustainability and ecological sustainability (Box 4). However, there are still concerns that the ecological element is not sufficiently rigorous. For instance while Naturland is reassured that the Lake Victoria Fisheries Organisation (LVFO) estimates that the stock is fully but not overexploited, there is in fact no agreed stock assessment and some studies have suggested that if fishing efforts continue at their current level stock collapse is a real possibility (MRAG, 2006).

### Box 4: Naturland certification pilot for Nile Perch, Tanzania, Lake Victoria

Naturland is currently piloting its standard on wild-caught fisheries with a Nile perch fishery in Tanzania (Bukoba) on Lake Victoria. In collaboration with the Netherlands importer Anova, and a local processor/exporter Vic fish, Naturland has developed specific project standards and plans to certify the fishery in 2009. The specific standards include ecological criteria on the state of the stock, fishing gear and data monitoring. Social critieria include labour standards, health and safety targets (e.g. provision of life jackets) and the socio-economic welfare of fishers' communities (e.g. access to health, primary education and financial services). There is also a trade element to the standard which requires the correct weighing of fish, transparent prices and direct access to processors rather than going through intermediary agents.

Some of the key limitations for developing countries to achieve certification are:

- **Lack of reliable scientific data:** as discussed above rigorous certification requires considerable scientific data to assess the sustainability of the stock. This can often be lacking in developing country fisheries, but new data-deficiency methods will help to address this.
- **Weak management:** certification is an important tool for incentivising sustainable fisheries, but it cannot alone ensure good management. The state and fishing industry or communities play a critical role, but developing countries often lack this capacity (UNEP, 2009).
- **The cost of certification:** is often a barrier for developing countries. For instance MSC certification can cost between \$20,000 and \$300,000 depending on the complexity of the fishery. Friend of the Sea certifications generally cost less. Development agencies and NGOs are increasingly filling this funding gap and helping developing country fisheries to achieve certification. For instance GTZ has supported the Naturland pilot, and WWF supported the MSC certification of the Mexican Rock Lobster and the Vietnam Ben Tre Fishery.
- **Lack of market demand:** while certification is popular in a number of markets, particularly northern European and US supermarkets, it does not have the same level of familiarity in southern European markets, which are often important destinations for developing country exports. For example a range of donors (e.g. GTZ and the World Bank) have been looking at opportunities to support MSC certification of the Senegalese artisanal octopus fisheries but the majority of this product is sold into Spain, France and

Italy, where there is little market demand for sustainability. Fish is also less packaged and more often sold in open markets or through independent fishmongers, offering fewer opportunities for using eco-labels (Montfort, 2007).

- **Multiple standards:** The range of certification schemes can be confusing for developing countries. While many of them claim to be fully in accordance with the FAO guidelines, there is currently no independent verification system that can confirm this (Globefish, 2008).

## Illegality/traceability schemes

Clearly, fish buyers are under increasing pressure to prove the legality of their supplies. This includes pressure from NGOs, but also pressure from governments. The EU, for instance, has drafted a proposed EU Council Regulation (2007/0223) which requests the flag state to prove the legality of fish entering EU ports. Many private sector organisations have made legality a key part of their fisheries procurement (e.g. Young's Seafood Ltd, an important fish processing company in the UK, see Box 5).

### Box 5: Young's ten principles for responsible fisheries procurement

1. **Legality:** only operate with suppliers that demonstrate compliance with relevant national and international law
2. **Objective assessment:** use of formal selection criteria
3. **Labelling:** communication on the origin of the fish
4. **Promotion:** actively promote use of certified sustainable species in the market place
5. **Continuous improvement:** encourage and work with suppliers to make improvements and minimise impacts on the environment
6. **Engagement:** engage in dialogue with a wide range of stakeholders
7. **Prohibition:** publicly state prohibited practices e.g. use of IUCN red listed species
8. **Research:** become involved with research projects into more effective technical conservation methods
9. **Traceability:** develop and improve traceability systems
10. **Ethics and environment:** Young's supports the ETI base Code of Conduct and the objectives of SA8000

Traceability schemes are seen as an important step in combating illegal supplies and ensuring that buyers know exactly where their fish is from. Some systems can be very sophisticated, for instance the South Georgian Patagonian toothfish is bar-coded at sea and each box can be tracked to its final destination. However there are also examples of lower-tech systems that have been applied in developing countries (Box 6). Certification schemes generally include chain of custody certifications to prove the labelled fish can be traced back to source. Traceability schemes, however, can also be applied to non-certified fisheries and help to confirm the origin and legality of the product.

### Box 6: Low-tech solution to traceability in Mauritania

Mauritania has recently set up a traceability system for artisanal pot-caught octopus that is destined for international markets. Boats are coloured to indicate which zones they fish in, and at the landing site each lot of fish that is transferred to the factory is recorded with information on the volume, date and areas of capture and the boats that contributed to the catch.

Source: Sid'ahmed Ould Abeid, National Fishermen's Federation, Mauritania

Traceability schemes are much easier to apply where the fishery concentrates on one species (rather than being a multi-species fishery) and where supply chains are vertically integrated, i.e. the links in the chain are relatively simple involving only a few actors. Figure 2 illustrates the different types of supply chains that can be present in fisheries and how traceability becomes more of a challenge where supply chains are more complex. Unfortunately, small-scale fisheries often have complex supply chains selling small volumes onto the open market or to a number of intermediaries.

In addition to traceability schemes, the private sector is looking at ways to reduce the risk of inadvertently buying illegal supplies. One mechanism being developed in the UK by

**Figure 2: Illustration of the range of supply chains existing within producing and importing countries**

| Production country | | Importing country |
|---|---|---|
| Fishermen deal in large volumes and have contracts with importers/retailers | Increasing complexity ↓ Declining traceability | Exporters have direct contracts with retailers |
| Fishermen have contracts with processors/exporters | | Exporters have direct contracts with buyers |
| Fishermen sell to a fixed trader | | Fish sold to fixed traders |
| Fishermen sell to a number of traders/intermediaries | | Fish sold via a number of traders |
| Fishermen deal in small volumes and sell at open market/local auction | | Fish traded on a wholesale market |

Seafish and the Food and Drink Federation (FDF) is a risk assessment tool to avoid any high risk products or producing countries. The aim is to develop an industry-led third-party auditable protocol that goes beyond the EU requirements and assesses and eliminates the risk of illegal products at each step in the supply chain.

The increased demand for traceability and proof of legality potentially adds additional challenges for developing countries, particularly the small-scale sector. In order to respond to the requirement for traceability, it is clear that supply chains for export will need to become more vertically integrated. The concern is that a lack of choice and competition in the market could reduce their negotiation capacity on trading terms and prices. Proving legality requires additional capacity on management but also documentation. While the EU has made the proof of legal fish mandatory for flag states, it does not provide any assistance to put these systems in place. This is in contrast to the forestry sector where the EU provides financial and capacity building support to assist developing countries tackle illegal timber through the Forest Law Enforcement, Governance and Trade Action Plan.

## Market Contribution to Poverty Reduction

Fisheries in developing countries provide a rich resource that often has the potential to generate economic wealth for the country, and benefit communities and individuals engaged in fishing. Industrial fisheries can be highly efficient and provide a high economic return. Conversely, small-scale fisheries can provide extensive employment, with around 26 million people in Africa, Asia and Latin America involved in marine and inland small-scale fisheries (FAO, 2005b).

However the full opportunity for fisheries to contribute to poverty reduction is often not realised. Small scale fisheries are faced with an array of problems including depletion of resources, lack of alternative livelihoods, migration of populations, displacement due to industrial or tourism development and conflicts with large commercial fishing operations (FAO, 2005b). Industrial fisheries face a different set of problems including the violation of human rights and labour codes on some trawlers and processing plants (particularly on illegal vessels or 'flags of convenience').

As illustrated above, the market can be used to encourage ecological sustainability and legality. There are also opportunities to use the market to encourage social and economic sustainability while also helping developing countries to export fish products and reach the required standards. These opportunities include setting relevant social and economic criteria, promoting innovative trading contracts and supporting organisational development.

### Social and economic criteria within supply chain mechanisms

One way forward is the inclusion of social and economic issues within certification schemes. Table 3 illustrates the type of issues included within organic, fisheries, aquaculture, coffee, forestry and Fairtrade sectors. The issues vary, but it is possible to break them down into the following categories: labour, community development and trade/economic issues.

**Table 3: Social and economic issues included within a range of certification schemes**

| | Organic | | Fisheries | | Aquaculture | | Coffee | | Forestry | | Fairtrade | |
|---|---|---|---|---|---|---|---|---|---|---|---|---|
| | Soil Association | Naturland | Fair Fish | Marine Stewardship Council | Friends of the Sea | Aquaculture Certification Council | Global GAP farmed shrimp | WWW Tilapia Dialogues | Starbucks Coffee | Utz-Kapeh Coffee | Forest Stewardship Council | Fairtrade standards |
| Labour issues | • | • | • | | | • | • | • | • | • | • | • |
| Socio-economic well-being of workers or communities | • | • | • | | | | • | • | • | • | | • |
| Development of democratic associations | • | • | • | | | | | | | | | • |
| Trade relations | • | • | • | | | | | | | | | • |
| Price premiums | | | • | | | | | | | | | • |

Labour issues, such as workers rights, minimum wage, and freedom of association, often form the core of social standards. For instance these are included within most aquaculture and coffee standards. There are also specific certification schemes that are based on the International Labour Organisation's (ILO) core standards (e.g. SA8001, BSCI and ETI). Labour issues are more generally applicable to larger scale operations dominated by the use of hired labour (i.e. industrial fishing); however it is possible to alter them to apply to smaller operations such as small scale or family run fisheries. For example the Naturland standard does not mention a ban on child labour but states that all children between 7 and 14 must attend school and not undertake any activities that could be hazardous to their health.

Criteria that are designed to encourage community development vary from respect of local community rights to actively promoting the socio-economic well-being of workers and of surrounding communities. For example in organic standards it is common to request the provision of certain social services for workers, such as access to health and education facilities. The Fairtrade standard supports the strengthening of community based organisations and ensures that the price premium is invested within local development projects.

One of the challenges in addressing social and economic issues within capture fisheries is that they are likely to vary considerably depending on the context, making it difficult to come up with a general standard. Naturland addresses this by developing a stakeholder round table for each fishery to develop specific standards and targets.

Economic criteria are quite rare and potentially more difficult to implement. For instance the Forest Stewardship Council sets down principles to support the 'efficient use of resources' and 'encouragement of local processing' without spelling out how this can be achieved. More concretely, some standards set down rules that govern the trading relations between the producers and the buyers. For instance the Soil Association and Naturland promote transparent pricing and direct links between producers and suppliers. Fairtrade has specific micro-economic criteria that set minimum prices and percentage price premiums to be paid to producers. However, no standards address the macro-economic issue of whether fisheries are economically sustainable.

Recent research has shown that the economic loss from inefficient fisheries over the past 30 years (1974–2007) amounts to $2.2 trillion (World Bank, 2008). Economic criteria within fisheries certification could help incentivise more efficient fisheries. While it is challenging to define such criteria, this could be done through governance indicators (Box 7) that could assess the institutional capacity to maximise economic returns from the fishery and ensure that the benefits are equally distributed (MRAG, 2008a).

### Box 7: Potential governance indicators

- Levels of participation and representation in management structures
- Accountability and transparency of management structures
- Transparency in the value chain
- Equitable allocation of access rights
- Effective provision of services
- Management maximises economic value of the resource

It is also possible to integrate social issues through business to business schemes. Private sector buyers or suppliers can request adherence to certain social standards. For instance labour standards such as SA8001 or ETI are often requested in the textile industry. Certain fish processors (e.g. Princess Tuna in the Maldives) have been applying such codes (in this case SA8001) to their processing plants.

## Responsible trade agreements

Traceability is now a key requirement and, to cope with this trend, supply chains need to become more vertically integrated. This presents opportunities to developing country fisheries in the form of higher prices (cutting out the intermediaries), investment into the socio-economic welfare of workers and communities and support to reach technical standards. The Fairtrade model also works on this basis by linking producers more directly with the end market and setting minimum prices that can help protect producers from the fluctuations of the market.

Even where minimum prices are not achievable or desirable, closer ties between the end market and the fishing producers mean that the importers or processors can invest in their suppliers, for example through supporting local development projects. Within the Naturland pilot, the exporter-processor has direct contracts with fishermen and therefore has a vested interest in supporting the socio-economic welfare of these communities. However, such contract arrangements can also turn exploitative if they are not accompanied by fair terms that allow for transparent pricing and equal negotiation powers. It is not always necessary to cut out the middlemen. As in the Vietnam Ben Tre Clam fishery, it is possible to bring traders into cooperative structures so that fishers and traders are working towards the same aims and the cooperative rules can govern fairness in the trading relations.

## Organisational development and rights-based management

Well defined access rights ('rights-based management') and strong local associations present a number of benefits for poverty reduction, but also for the efficiency of supply chains. Organisational development improves the negotiation capacity of producers in defining fair trade arrangements, but also with government to demand services such as access to market infrastructure, health and education. Access to credit is important in allowing fishermen to be independent, thereby increasing their negotiating power. An interesting observation is that the developing country fisheries that have achieved MSC certification to date have clearly defined access rights and either effective state management in place for industrial fisheries (e.g. South African hake fishery) or strong local associations in the case of small scale fisheries (Box 8 and Box 9).

### Box 8: MSC certification of the Mexican Rock Lobster

The Baja Californian Mexican Rock Lobster gained MSC certification in 2004 and was re-certified in 2007. The fishery is located in north-west Mexico and involves 1,159 fishers. It is organised by a system of cooperatives, and access to the fishery is granted through long-term concessions (20 year periods). The cooperatives are responsible for the production, processing and marketing of the lobsters, and also fulfil a social role by providing wages to the fishermen and using profits to invest in social infrastructure such as electricity, road, water, and services such as schools and clinics. The eco-label has not further improved ecological sustainability but has rewarded existing good practices. However, one of the major benefits has been the increased likelihood that the cooperatives will be re-issued with concessions. The cooperatives have a regional representation that can engage with the state on development issues. Certification is also though to have assisted in the granting of a 20-year request to federal government to provide electricity, improve the road infrastructure and improve processing infrastructure.

Source: Venkataraman and Shabelman (2008)

## Box 9: Vietnam Ben Tre Clam Fisheries

The Ben Tre Clam fishery is located in the south-east of Vietnam, and involves three coastal districts where fishermen hand-gather clams. The fishery was originally identified by WWF's community fisheries programme as being potentially suitable for certification. The fishery is currently being assessed using the GASSDD methodology and a decision is due in 2009. The introduction of cooperatives and allocated fishing rights has been very successful in terms of fisheries management and sustainability, and migrant fishermen have effectively been excluded from the fishery. As well as sustainable management, the supply chain relations in this fishery also suit certification. The cooperative structure includes both representatives of fishers, but also traders, which assists in obtaining high quality standards and traceability. Although not required for the MSC certification, this also means that fishers and traders are working towards the same aims rather than against each other and fairer trading relations emerge.

Rights-based management also has the advantage of reducing over-exploitation and improving the economic performance of the fishery. If fisheries are managed to maximise economic returns it is possible for the government to extract rents and reinvest these into development, thereby further reducing poverty (Cunningham and Neiland, 2005). Well defined access rights promote the sustainability of resources by giving an incentive to the owners of the rights to maintain productivity, and with the right legal framework and enforcement this can also reduce illegal fishing. However, there are challenges in allocating the rights. Senegal has been discussing the idea of fishing permits for the octopus fishery since 2000, but the political nature of restricting access and turning some people away from the fishery has been too delicate to take action.

## Conclusion and Recommendations

Developing countries play an important role in fish trade, but there are opportunities to increase this further. Supply chains are playing a pivotal role in driving sustainability and legality forward, but could go further to assist developing country involvement and help reduce poverty in the fisheries sector. International agencies and donor organisations supportive of developing country fisheries (e.g. FAO, DFID, GTZ, OECD, World Bank, UNEP and the Commonwealth Fisheries Programme) can support this progression.

- **Address constraints to trade and certification**

There are still a large number of constraints to developing country fisheries engaging in international trade. Quality and hygiene are the first issues that have to be addressed. Moving on from this, fisheries need support to demonstrate traceability and sustainable management. Both small scale and industrial fisheries in developing countries, and their management structures, need support to cover the costs and develop capacity to reach certification and private standards.

- **Integrate social and economic issues into certification and procurement criteria**

By including criteria that support sustainable development, certification and business to business schemes have the potential to promote long-term benefits at the community and national level. This needs to be criteria that cover economic and social objectives as well as ecological sustainability. NGOs can also play a role by highlighting the social and poverty issues associated with fishing rather than focusing only on environmental issues. There is a need for guidance on how to address social and economic issues in fisheries certification either through specific criteria or governance indicators.

- **Support responsible approaches to vertical integration through fair trading contracts**

A key way in which trade in fisheries products can contribute to poverty reduction is through favourable trading terms. This can include linking producers more directly with the end market to increase the prices they receive, linking traders and fishers through cooperative structures with common aims, promoting transparency in trading agreements and making trade conditional on investment in local development or access to services (e.g. credit).

- **Support capacity building for rights-based management**

Rights based management approaches and wealth-based fisheries management can be piloted and supported to benefit sustainability and reduce illegal fishing, but also to promote economic returns. Combined with strong and democratic associations, this approach can also support the equitable distribution of benefits. Cooperative structures (rather than individual fishermen) assist with traceability, covering the costs of certification and increasing the voice of fishing communities with the government. As in the forestry sector, capacity building to improve governance and reduce illegal production requires investment and resources.

## References

Cunningham, S. and Neiland, A.E. (2005) *The Role of Fisheries in Economic Growth and Poverty Alleviation: Towards a Wealth-Based Approach to Fisheries Management*. Key-sheet prepared for the UK Department for International Development (DFID).

FAO (2002) *Implementation of the International Plan of Action to prevent, deter and eliminate illegal, unreported and unregulated fishing*, FAO Technical guide for responsible fisheries (Rome: FAO).

FAO (2003) *Report of the Expert Consultation on International Fish Trade and Food Security Casablanca, Morocco*, 27–30 January 2003. FAO Fisheries Report. No. 708 (Rome: FAO).

FAO (2005a) *Guidelines for the eco-labelling of fish and fishery products from marine capture fisheries* (Rome: FAO).

FAO (2005b) *Increasing the contribution of small-scale fisheries to poverty alleviation and food security* (Rome: FAO).

FAO (2007a) *Draft Technical Guidelines on Responsible Fish Trade, Technical Consultation on International Guidelines for Responsible Fish Trade* (Rome: FAO).

FAO (2007b) *The State of World Fisheries and Aquaculture*, 2006 (Rome: FAO).

FAO (2008) *Status and Importance of Recent Events Concerning International Trade in Fishery Products. Committee on Fisheries: Sub-committee on fish trade*. Report from the Eleventh Session Bremen, Germany, 2–6 June 200.

Globefish (2008) *Ecolabels and Marine Capture Fisheries: Current Practices and Emerging Issues*. Globefish Research Programme, 91.

Keizire, B.B. (2004) *A Case Study for Uganda. Policy Research: Implications of Liberalisation of Fish Trade for Developing countries* (Rome: FAO).

Montfort, M.C (2008) *Market Trends in Seafood Trade: Europe. Presentation at the Seafood Choices Alliance Summit*, Barcelona, 28–29 January, 2008.

MRAG (2006) *Strengthening of the Fisheries Monitoring Programme and Development of a Fisheries Management Decision Support Tool (FMDST) for Lake Victoria*. Consultancy Report No. 26 for the EU Project: Implementation of a Fisheries Management Plan for Lake Victoria.

MRAG (2008a) *Ethical sourcing of wild-caught fish from developing countries: Opportunities and Constraints*. Discussion Paper prepared for a workshop at DFID, 29 May 2008.

MRAG (2008b) *Briefing Paper: Fisheries Product Strategy*. Commissioned work by the Fairtrade Foundation.

MSC (2002) *MSC Principles and Criteria for Sustainable Fishing*.

Naturland (2008) Naturland Association Standard Committee reply on the Specific Standards proposed by the Standard Round Table during the 2nd meeting in Bukoba Kolping Formation Center Tanzania, 14–15 April 2008.

UNEP (2009) *Challenges for the Sustainable Consumption and Production of Fisheries Products: ecolabelling, certification, and other supply chain issues*. Chair's Summary of meeting: Paris, 18–19 September 2008.

Venkataraman, B. and Shabelman, A. (2008) *The Political Implications of Fisheries Certification: Lessons from Baja California Sur. A Policy Analysis Exercise (PAE) on the Baja California Sur Lobster Fishing Cooperatives and Marine Stewardship Council Certification*. Commissioned by MSC.

World Bank (2008) *The Sunken Billions. The Economic Justification for Fisheries Reform*. Agriculture and Rural Development: Sustainable Development Network (Rome: FAO and Washington: World Bank).

## Notes

1 WWF has developed a series of seafood guides for different markets including France, Hong Kong, Germany, Spain, Finland, Denmark, Norway, Switzerland, Netherlands, Sweden, Poland, Belgium, Indonesia and South Africa. WWF make use of the Seafood Choice Alliance methodology for assessing wild-fisheries which was circulated for consultation in August 2007.

2 Mahi mahi from Ecuador and Peru are going through the MSC trial assessment for data-deficient fisheries.

# 11 Customary Law and Inshore Fisheries in Pacific Island States

Yoli Tom'tavala

## Introduction

This chapter examines the implications of customary law to sustainable inshore fisheries in Pacific island countries, especially in those that are members of the Commonwealth.[1] This issue arises because the lives of most indigenous Pacific islanders are ordered in part by customary laws. With the exception of those who live in the highlands of Papua New Guinea, the majority of Pacific islanders live along coastlines and have therefore evolved livelihoods and cultures that are intricately connected to the sea and its resources.

Today, these indigenous communities are subsumed within the boundaries of modern, independent states which are governed by introduced political and legal systems. Hence, their lives are regulated by two distinct politico-legal systems: on one hand, the Westminster inspired system and, on the other, the customary systems of social organisation and livelihood. The former system is more dominant in towns, cities and localities that are within easy reach of governmental services. In the far-off villages, islands and atolls where the reach of the government is limited, customary systems continue to dominate affairs within the communities concerned.

Although the state and its laws have a say on the affairs of people in remote communities, in practice, their everyday existence is plainly ordered by customary norms and practices. This observation applies even to the inshore waters and resources where customs continue to determine such aspects as who owns or claims these areas, who can access the resources of these areas, under what circumstances and what penalties should be imposed when people fail to obey customary norms and strictures.

In order to answer the issue that was stated earlier, the paper canvasses the following matters. It begins the substantive discussion by making some general observations about the inshore domain and its significance to the social, cultural and economic lives of Pacific islanders. It then discusses in broad terms the notion of 'customary laws' of indigenous Pacific islanders and examines their interface with official legal systems of Pacific island states. It follows this by analysing the implications of customary systems on sustainability of inshore areas and postulates some recommendations for future action.

## Inshore Domain in Pacific Island States

For the purposes of this paper, the term 'inshore domain' refers to beaches, deltas, estuaries, lagoons, reefs, the living and non-living resources thereof and the surrounding waters that are traditionally claimed or used by indigenous communities. The distance from shore can range from one kilometre to tens of kilometres depending on the nature of the coastal environment and the extent of its use by indigenous coastal communities.

The nature of an inshore domain and its productivity differs from country to country and even from one coastal location to another, depending on the geographical configuration of each island concerned. Broadly, three main types of coastlines dominate in the tropical Pacific.

The first consists of coasts which abut either narrow fringing reefs or cliff faces which drop off into deep blue seas. These are found around raised single-island states such as Nauru and Niue but also along portions of bigger islands in other Pacific island states where there are high wave energy coasts. The sandy coasts, narrow reefs and violent wave action contrive to ensure that marine life forms are least abundant in these localities. Nevertheless, the fringing reefs contain many types of shellfish, crustaceans and swimming species of fish which coastal communities rely on from time to time. The nearby waters are also fished with hand lines for both pelagic and bottom fishes.

The second consists of atolls or rings of reefs which surround lagoons where the waters are relatively shallow and calm and the seabed is mainly covered by extensive beds of sea grass. These are common in Kiribati, Tuvalu and smaller island groups in the Solomon Islands, PNG and Fiji. The flora and fauna of these are more varied and abundant because the sheltered nature of the coast affords marine life ideal conditions to spawn and aggregate. Hence, shellfish, crustaceans, swimming species, seaweeds, turtles, sea urchins and dugongs abound and are quite intensively utilised by nearby coastal communities who utilise different fishing methods.

The third consists of estuaries where freshwater bodies such as lakes adjoin the coast or where streams or rivers flow into the sea. Many estuarine areas are also covered by mangroves and swamps. This type of coastline is found along portions of coasts of the larger Melanesian and Polynesian islands which have significant aquatic water bodies. These afford the most productive areas for all manner of marine life because the runoff from rivers and streams provides constant supply of food for plants and animals. Furthermore, mangroves afford excellent spawning grounds for all kinds of marine species.

Inshore domains and their resources have historically provided the mainstay of the socio-cultural and economic livelihoods of coastal communities. After all, it was through sea voyages that Pacific islanders set out to discover and settle the islands which they now inhabit. In their new settlements, the seas continued to provide them with food, water to bath with, salt to cook or preserve their food, material to use in construction and a myriad of other uses. But the relationships of the islanders with the seas and marine life extend even to the metaphysical realm. As a commentator correctly wrote of Oceanic people "In their world all things were connected; fish might be relatives as well as food, and Gods and ancestors could take multiple forms".[2] Thus some communities claim family ties with certain marine species whilst other species are reserved for the exclusive use of chiefs or persons of superior rank. It is for these reasons that Pacific islanders have complex perceptions over their inshore environments which transcend the notions of property or sovereignty that the legal systems postulate.

Without doubt the most significant activity in the inshore domain is fishing. As one commentator has correctly observed, "the people of the Pacific islands have been almost totally and continuously reliant on their inshore fisheries for hundreds, or in some cases thousands, of years ..."[3] Today the inshore waters support three main forms of fisheries activities: subsistence, artisanal and commercial.

Subsistence fishery refers to situations where the catch is mainly consumed by fisher families or is bartered for agricultural commodities or is given away because of customary, church or other commitments. Artisanal fishery involves situations where the catch is mainly sold in local markets but part of which is normally retained for the fisher's personal consumption. According to the Secretariat of the Pacific Commission, "*Per capita* seafood consumption in some remote atolls is over 250 kg annually, whilst other agriculturally oriented PICTS consume over 50 kg of marine protein".[4] Both forms of fishery are undertaken mainly by Pacific islanders who employ a wide range of fishing methods and gear such as fishing hand lines, push nets, spears and collection. The main targeted species include reef fishes, pelagic species, shell fish, seaweeds and crustaceans. It has been estimated that in the 1990s these sectors yielded around 91,000 metric tonnes which had an estimated value of around USD$94,800,000.[5]

The third form of fishery is commercial fishing where the sole aim is to harvest for export purposes. Targeted species include bêche-de-mer, trochus, mother-of-pearl shells, dried shark fins, live reef fish, seaweeds, giant clams and corals. It has been estimated that this fishery yielded around 38,500 metric tonnes which were worth around USD$92,800,000.[6]

One of the biggest problems facing inshore fisheries is best stated by Adams and Ledua (1997) who wrote:

> *[P]olicy-makers do not really know the status of inshore living marine resources in most parts of the Pacific Islands region. We do not even know the extent of their usage by Pacific Island societies, with any degree of accuracy ... Just what level of economic activity is possible for different fisheries? What is the current state for exploitation? What is the sustainable level of exploitation? ... Who fishes? What percentage of the catch enters the cash economy?*[7]

This much was acknowledged in principle 1 of the *Pacific Islands Regional Coastal Fisheries Management Policy and Strategic Actions (Apia Policy) 2008–2013* which calls for "improving our understanding of important fisheries species and the ecosystems on which they depend" (in part) by "collation of knowledge on the life histories of commonly harvested species and their ecosystems".

What is not at issue is that stocks of inshore living resources are declining and some commercially-exploitable stocks are in serious danger of being completely wiped out. No doubt these are due mainly to increased fishing efforts that are wrought by increasing populations and demands for cash in order to meet school fees and other financial commitments. This can also be attributed to alterations in ecosystems that are caused by environmental factors such as "presence of sewage and nutrients in coastal waters, garbage dumps located at the sea edge, excessive coastal developments, silt entering the sea, and loss of beaches due to sand mining".[8]

## Customary Laws and the Legal Systems of Pacific Island Countries

Prior to contact, customs and culture provided the sole basis of the indigenous Melanesian, Micronesian and Polynesian peoples' claims to, uses of and husbandry of their natural sources, including the resources of the inshore domain. Even though colonialism and modernism have lessened the impact of these in some segments of Pacific island communities, in others (such as remote villages, islands and atolls) customary systems continue to dominate affairs within the communities concerned. The strength of their customary systems leads one to conclude that these societies are still governed by customary laws.

What are customary laws? These are part of the cultures of indigenous Pacific islanders. They consist of rules, practices, beliefs, values and processes about every aspect of their lives, including their natural environment and the use of its resources for food, material implements and other uses. The content of these are shaped partly by traditions and by evolving experiences such as those that are being brought about by religion, politics and formal laws. Hence, by their nature customary laws are evolving. The bulk of customary laws are unwritten and are passed along from generation to generation through word of mouth, observation and practice and, in some cases, special training through initiation or similar ceremonies of passage. Nowadays there are attempts in a few places to put in writing some rules of customary law.

It remains to analyse the various implications of the customary systems on sustainable inshore fisheries. For the time being, it suffices to consider some basic aspects of customary legal systems that have a bearing on the Pacific islanders' claims to, governance of and usages of inshore domains and fishery resources. These are: (a) customary land ownership, (b) roles of customary authorities, and (c) economic lifestyle.

### Rights to customary land

Thankfully, alienation of land from indigenous owners was not one of the hallmarks of colonialism in Pacific island states. Instead, apart from the few instances where land was alienated for public purposes and economic development, most of the land remained in the hands of customary owners. With the exception of Tonga where the 1875 Constitution vested ownership of all land in the King, in all other Pacific island states the colonial and post-colonial laws recognised the rights of indigenous Pacific islanders to their lands which, in most cases, were to be held according to customary tenure rules.

The significance of this is that according to customary laws of indigenous Pacific islanders the notion of 'land' extends from *terra firma* down to beaches, the inter-tidal zones, contiguous estuaries, reefs or lagoons and the adjacent waters which they plied or with which they have some practical or metaphysical connections.[9] Consequently local indigenous landowners claim various rights to inshore domains, fishery and other resources of these areas. These customary rights form the basis of customary marine tenure systems where the communities concerned have established rules or practices regarding ownership of, access to, use and control of these areas and their resources. According to the *Pacific Islands Regional Coastal Fisheries Management Policy and Strategic Actions (Apia Policy)*

*2008–2013* "the concept of customary ownership of (marine) resources is common in most PICTs with the exception of Tonga and Kiribati ...". The latter part of this section will show that this situation is exacerbated in most countries because the law is uncertain as to who really owns the inshore waters.

## Traditional roles of customary authorities

Because inshore domains and resources are regarded as part of the land, it followed that clans, tribes, families or villagers which own the adjacent land claim ownership or similar rights and play prominent roles in managing and controlling access to and uses of the inshore areas and their resources. In practice the management and control functions lay in the hands of select individuals such as family heads, elders or chiefs who invariably dominate decision-making in community affairs. Their traditional powers include making decisions about taboo sites, placement of *namele* leaves or other taboo signs to conserve resources, settling disputes amongst members of the community and deciding on matters concerning the general welfare of their people.

The effectiveness of these traditional authorities depends on the extent to which they are able to sanction their decisions or are respected and obeyed by other members of the local community. It is accepted throughout the Pacific that chiefs and other traditional authorities are not always effective as decision-makers today because they either lose the ability to sanction or are not respected by members of younger generations who are schooled in notions of equality, human rights and democracy. Another negative aspect is that most chiefs and other traditional authorities do not have customary rights to control actions of people who do not come under their traditional jurisdiction, so their effectiveness is limited.

## Economic Lifestyle

As may be implicit from earlier discussions, the majority of indigenous Pacific islanders directly depend on the seas for food and to make money from the sale of fish and other marine products. The subsistence and artisanal fisheries that Pacific islanders engage in means that in most cases they would place their own needs and interests before any competing demands that states are likely to impose such as through the form of licences issued to commercial fishermen or measures to ensure proper conservation and management of stocks. In any case the ability of fisheries officials to enforce licenses and management and conservation measures in remote locations is limited so in many places the needs of local communities hold sway in inshore waters. Where local populations are large and the level of fishing effort is intensive, this scenario easily leads to over-exploitation of stocks.

Another implication of this scenario is that the introduction of any conservation or no-take measures is likely to fail because it goes against the very essence of subsistence lifestyles. Can indigenous fisher folk be reasonably expected to abide by conservation rules prohibiting taking of resources if the children are craving for animal protein? Or where money is needed to pay for children's school fees? After all, if they cannot meet their needs from their surrounding environment and natural resources, where else can they get help? These rhetorical questions pose clearly the dilemma that confronts national inshore fisheries managers.

What place do customary laws have within the legal systems of Pacific island states? We now consider this question by examining, firstly, how the legal systems in Pacific island states interface with customary laws in general and, secondly, with the customary interests of indigenous right holders to inshore domains and fishery resources thereof. In fact, this question was raised by the Sixth Technical Meeting of the Forum Fisheries Committee way back in April 1992 which led to the engagement of a consultant who drafted a report entitled *An Overview of Constitutional and Legal Provisions Relevant to Customary Marine Tenure and Management Systems in the South Pacific*, FFA Report 93/23 (1993).[10] Hence, in discussing these two issues I rely a lot on the analysis by Ms Pulea but have added in aspects that she missed or which have evolved since the publication of the report in 1993.

## Legal Recognition of Customary Law in General

The contemporary legal systems of Pacific island states are essentially dualistic because customary laws are practised in the face of established constitutional, statutory and other introduced types of laws. Section II of the Pulea report has canvassed the framework of laws in Pacific island states which places constitutions first followed by statutory laws, case laws, and (in some countries) customary laws. Principles of English common law and equity are also recognised but their application is subject to the proviso that they are not inappropriate to local circumstances or are not inconsistent with customary laws.

Customary laws are codified or recognised for various specific matters such as customary land or as an issue in criminal sentences. As a source of law of wider application, these are recognised by the constitutions or statutory laws of the following member countries of the Commonwealth. In essence three approaches are used as summarised below.

- The *first* involves constitutional or statutory recognition of customs as a source of national laws where the recognising provision is self-executing. This is done in Nauru, PNG, Solomon Island and Vanuatu.
- The *second* involves constitutional recognition of custom as a source of laws but to be prescribed by enabling legislation or case law. This was followed by the Cook Islands, Fiji Islands and Samoa.
- The *third* is where legislation has provided for the application of customs related to specified subject matters in civil and criminal proceedings before the formal courts. This is done in PNG, Solomon Islands, Kiribati and Tuvalu.

## Cook Islands

Article 66A of the *Cook Islands Constitution 1965* says that Parliament may make laws recognizing or giving effect to custom and usage having particular regard to the customs, traditions, usages and values of the indigenous people of the Cook Islands. Until such time as an Act otherwise provides, custom and usage shall have effect as part of the law of Cook Islands, as long as these are not inconsistent with a provision of the Constitution or any enactment.

## Fiji Islands

Similarly, section 186 of the Fiji Islands *Constitution (Amendment) Act* 1997 authorises the "Parliament to make provision for the application of customary laws and for dispute resolution in accordance with traditional Fijian processes". In doing so, the Parliament must have "regard to the customs, traditions, usages, values and aspirations of the Fijian and Rotuman people".

## Kiribati

Section 4 of the *Laws of Kiribati Act* 1989 lists customary law as one of the laws of Kiribati and adds in section 5 that these "shall have effect as part of the law of Kiribati except to the extent that it is inconsistent with an enactment or an applied law".

## Nauru

Section 3 of the *Customs and Adopted Laws Act* 1971 of Nauru says that "the institutions, custom and usages of the Nauruans ... shall be accorded recognition by every court and shall have full force and effect as law".

## Papua New Guinea

Section 9 of the *Constitution* 1975 lists 'underlying law' as a source of laws to be provided for by statute. The envisaged act was the *Underlying Law Act* 2000 which provides in section 3(1)(a) that the underlying law includes 'customary law'.

## Samoa

Section 111(1) of the *Samoan Constitution* 1962 defines 'Law' to include "any custom or usage which has acquired the force of law in Western Samoa or any part thereof under the provisions of any Act or under a judgment of a Court of competent jurisdiction".

## Solomon Islands

Schedule 3(1) of the *Constitution* 1978 states that 'customary law shall have effect as part of the law of Solomon Islands'. It also authorises the making of an act to regulate the proof and pleading of customary law and for related purposes. The Act of Parliament envisaged here is the *Customs Recognition Act* 2000 which is essentially based on the PNG *Customs Recognition Act* Cap 19 (Hviding, 1989).

### Tuvalu

Like the Kiribati law, section 4 of the *Laws of Tuvalu Act* 1987 lists customary law as one of the laws of Tuvalu and adds in section 5 that these "shall have effect as part of the law of Tuvalu except to the extent that it is inconsistent with an enactment or an applied law".

### Vanuatu

Article 95(3) of the Vanuatu *Constitution* 1980 states simply that "Customary law shall continue to have effect as part of the law of the Republic of Vanuatu".

Despite the evident widespread recognition of customary laws, the reality is that these laws are essentially rhetorical because in practice customary laws are given little or no practical recognition in the affairs of modern Pacific island states.

## Legal Recognition of Customary Marine Interest and Tenure Systems

Unlike customary lands, which have received quite widespread legal recognition and protection, the situation concerning customary rights to marine territory is quite different. This stems from two related effects of the common law system which Pacific island states have followed. The first is that the introduced common law makes a conceptual distinction between *terra firma* and maritime domains so ownership of interests in coastal lands does not amount to ownership of adjacent maritime areas. Instead, at common law the ownership of the inter-tidal lands, the adjacent belt of seas and seabed and subsoil thereof were vested in the Crown. This being so, colonial and post-colonial officials were not inclined to recognise the full spectre of customary marine rights.

During the colonial period, statutes expressly extinguished customary rights and vested them in the Crown in some dependencies. For example, the Samoa *Constitution Act* 1920 (NZ) vested in the Crown all land between the high and low water marks and all waters within the territorial limits. Section 101 of the 1968 Constitution of Samoa declares same as 'public land' and are vested in the state. The *Cook Islands Act* 1915 (NZ) simply stated that customary titles to land do not extend beyond the high water mark because such land belonged to the Crown (s.419). In the British Solomon Islands Protectorate, the *Land and Titles (Amendment) Ordinance* 1964 similarly vested the foreshores and seabed and subsoil in the Commissioner of Lands so these became public lands.

In all the other Pacific islands, the reluctance of the colonial administrators to recognise fully customary marine territorial rights can only be attributed to the common law doctrine of Crown prerogative (above). However, it has been established by case authority that this doctrine is subject to local customs such as exclusive fishing rights where such rights were continuously observed since advent of colonisation, certain in principle and application, and reasonable at the time of its conception.[11] More tellingly, in countries like Kiribati, PNG, Solomon Islands and Tuvalu where customary law is stated to override common law, I submit that where there are no statutory provisions to the contrary then customary rights to inshore marine territory including reefs and waters should override the doctrine of Crown prerogative over the foreshore areas.

As things stand, Pacific island states provide for very minimal recognition of customary marine rights. In Kiribati, PNG, Solomon Islands and Tuvalu the civil courts are authorised to apply and enforce customary laws such as those concerning "the ownership by custom of rights in, over or in connection with any sea or lagoon area, inland waters or foreshore or reef, or in or on the seabed, including rights of navigation or fishing; (or) the ownership by custom of water, or of rights in, over or to water …"[12] In Fiji the *Fisheries Act* Cap 158 makes it an offence for an outsider to take fish or shellfish on any reef or seabed "in any area in respect of which the rights of any mataqali or other division or subdivision of the Fijian people have been registered by the Native Fisheries Commission in the Register of Native Customary Fishing Rights …" (s.13). In PNG, the *Fisheries Management Act* 1998 says that "The rights of the customary owners of fisheries resources and fishing rights shall be fully recognised and respected in all transactions affecting the resource or the area in which the right operates". The scope of this right is not totally clear but in practice indigenous right holders are usually paid royalties for bait fish harvests in waters that they customarily claim. In Samoa, the Village Fono or Council in coastal villagers are reportedly empowered to adopt bylaws for the purpose of ensuring the conservation of fishery resources.[13] On a similar vein the *Falekaupule Act* 1997 authorises each Kaupule or island council in Tuvalu, amongst other things, to "ensure the proper management and use of the physical and natural resources in the Falekaupule area and to provide for the improvement and control of fishing and related industries in accordance with the Fisheries Act" (s.40 and sch. 3).

## Implications of Customary Law on Sustainable Inshore Fisheries

In most countries of the region, indigenous Pacific islanders assert proprietary and usage rights to inshore domains and resources. They also rely on the fishery and other resources for their subsistence and artisanal livelihoods. Access to and use of these resources is based on customary and evolving tenure rules. In all countries, the traditional command-and-control approach to marine management has been ineffective in these places either because the resources of the state required to enforce laws are limited or because the imposed measures contradict customary rights.

Consequently, customary law and customary marine tenure systems offer a viable alternative approach to sustainable inshore fisheries management especially in places where chiefs or other customary authorities are still respected and whose decisions are enforced. In some places, such as in some Vanuatu communities,[14] this is done under the sole authority of customary law. In others such as Samoa and Tuvalu the marine resource management roles of traditional authorities is strengthened by laws which give them power to make bylaws that can be enforced through the courts.

The flipside is that the tenure rules and practices of many customary marine right-holders allow for fairly open access where increasing populations and increased demands for cash can lead to unsustainable fishing efforts and depletion of stock. In such cases customs, laws and science should be integrated so that locals continue to be involved on property or rights-based management or increased stakeholder participation regimes; albeit they do so under the enabling of laws that should reflect sound scientific principles. The similarity between customary marine rights and emerging mechanisms such as Community-Based Fisheries Management (CBFM), Community-Based Marine Resources Management (CBMRM), Locally-Managed Marine Area (LLMA) network, Marine Protected Areas (MPAs) and Conservation Areas (CAs) explains why these initiatives are gaining increasing popularity in Pacific island states.

These are working well because local communities appreciate their economic interests which then motivate them to take part in sustainability measures. The problem is, apart from a few instances, most of these are essentially voluntary schemes so that their legal status and the legal enforceability of their rules remains suspect. Thus it is imperative that these measures be founded on sound legal and policy frameworks.[15] At the same time, the plurality of natural resource governance in Pacific island states requires that both customary and state authority and norms should be integrated.[16]

As a case in point, I discovered that the management and operation of the much touted Vathe Conservation Area (CA) in Vanuatu was effectively in disarray after foreign donors and consultants had left. This was because although the concept was provided for in legislation, the Act did not provide for legal enforcement of the rules of the CA. Instead these were left to be enforced through the customary office of the kastom chiefs. Unfortunately, the paramount chief was not included in the management committee and showed his displeasure by openly transgressing the CA rules. Because the courts had no jurisdiction to try him, his actions led others to question the integrity of the rules and their own participation in the venture. What became quite obvious is that the consultants went to great lengths to entrench participatory approaches by ensuring representation of women, youth and others in the management committee but omitted the paramount chief. This failure became the Achilles heel of the project. Another weak point was that the legislation failed to enable breaches to be enforced in the courts perhaps relying upon the common view that chiefs effectively settle disputes in their communities. In this locality, the paramount chief's power and influence was in decline so he relocated away from the village and did not have much to do with dispute settlement there.

The moral to this story is that when it comes to customary issues it is wise to avoid making generalisations. Instead, local priorities and situations should be considered and dealt with on their own merits. For this reason, it is not feasible to devise and introduce 'model laws' for use in all Pacific island states. Instead sustainable inshore fisheries mechanisms should be devised to fit local circumstances incorporating law and customs.

## References

Adams, Tim and E. Ledua (1997) *Inshore Resources Management and Conservation: Current Trends and Alternate Strategies*, Paper prepared for the Pacific Science Association Inter-Congress Meeting (Marine Resources Symposium) (Suva).

Craig, D., E. Kwa, M. Jeffery, Y. Tom'tavala and J. Rose (2008) *Indigenous Governance of Natural*

*Resources in Melanesia: a Project on Harmonising Customary Institutions, State Laws and Environmental Management Approaches* (Palo Alto, CA: Christensen Fund).

D'Arcy, Paul (2006) *The People of the Sea: Environment, Identity and History in Oceania* (Hawaii: University of Hawaii Press).

Fa'asili, U. and A. Mulipola (1999) *The Use of Village By-laws in Marine Conservation and Fisheries Management* First Heads of Fisheries Meeting, Noumea, New Caledonia, 9–13 August 1999.

Gillet, R. and C. Lightfoot (2001) *The Contribution of Fisheries to the Economies of Pacific Island States* (Asian Development Bank Pacific Study Series).

Hickey, F.R. (2006) Traditional Marine Resource Management in Vanuatu: Acknowledging, Supporting and Strengthening Indigenous Management Systems *SPC Traditional Marine Resource Management and Knowledge Information Bulletin* # 20, pp. 11–23.

Hickey, F.R. and R.E. Johannes (2002) 'Recent Evolution of Village-based Marine Resource Management in Vanuatu' *SPC Traditional Marine Resource Management and Knowledge Information Bulletin* # 14, pp. 8–21.

Hviding, Edvard (1989) *All Things in Our Sea: The Dynamics of Customary Marine Tenure, Marovo Lagoon, Solomon Islands* (The National Research Institute Special Publication No. 13, Papua New Guinea: 1989).

Johannes, R.E. (2002) 'The Renaissance of Community-based Marine Resource Management in Oceania' *Annual Review of Ecological Systems*, pp. 317–340.

Kabui, F. (1997) Crown Ownership of Foreshores and Seabed in Solomon Islands'(1997) *Journal of Pacific Studies* 21, pp. 123–144.

Kuemlangan and H. Teigene (Undated) *An Overview of Legal Issues and Broad Legislative Considerations for Community Based Fisheries Management* (Rome: FAO Legal Office, Rome).

Secretariat of the Pacific Community, *Pacific Islands Regional Coastal Fisheries Management Policy and Strategic Actions (Apia Policy)* 2008–2013.

## Notes

1 These consist of Cook Islands, Fiji, Kiribati, Nauru, Niue, Papua New Guinea (PNG), Samoa, Solomon Islands, Tonga, Tuvalu and Vanuatu. However, the paucity of relevant information relating to some of these countries means that some will receive more attention in this piece than others.

2 D'Arcy (2006), p. 40.

3 Adams and Ledua, (1997), p. 3.

4 In *Pacific Islands Regional Coastal Fisheries Management Policy and Strategic Actions (Apia Policy)* 2008–2013, slide # 2. By comparison the annual consumption per capita in continental European states is 8 kg. 'PICTS' means 'Pacific Island Countries and Territories'.

5 Gillet and Lightfoot (2001) pp. 35–6. The authors used figures from the late 1990s so, arguably, both the quantum and value have increased in the intervening years.

6 Above; as noted earlier these calculations used figures from the late 1990s so both the quantum and value of catch should have increased in the intervening years.

7 Note 3, above.

8 Note 4, slide # 16.

9 Baines similarly observed as follows: 'Adjacent reefs and intervening lagoons, mangroves and estuaries are seen as integral components of that land, not as distinct entities separated from land from a certain tidal level'. Quoted by Pulea in *An Overview of Constitutional and Legal Provisions Relevant*

*to Customary Marine Tenure and Management Systems in the South Pacific*, FFA Report 93/23 (1993) at p. 18.

10 By Mere Pulea, whose report covered quite a broad spectrum of related issues such as the framework legal systems in Pacific Island countries, customary rights and marine tenure, legal arrangements in the marine environment, arrangements for management and conservation of resources and suggested principles for integration of customary principles in legal systems.

11 In the case of *Hanasaki v. O.J. Symes* as discussed by Kabui, 'Crown Ownership of Foreshores and Seabed in Solomon Islands' 21 *Journal of Pacific Studies* (1997) 123–144.

12 This provision is similar in the laws of all four countries but has not received the benefit of any judicial interpretation in any of them. However, its terms appear unambiguous.

13 Fa'asili and Mulipola, 'The use of Village By-laws in Marine Conservation and Fisheries Management' First Heads of Fisheries Meeting, Noumea, New Caledonia, 9–13 August 1999.

14 See Hickey, F.R. (2006) and Hickey, F.R. and R.E. Johannes (2002).

15 For similar sentiments see Kuemlangan and Teigene (undated).

16 Craig, Kwa, Jeffery, Tom'tavala and Rose (2008), p. 2225.

# 12 Impact of Climate Change on Marine Fisheries in the Commonwealth Caribbean

Arthur C. Potts and Judy Rocke

## Introduction

Fisheries are affected by a wide range of human, physical and biological factors. The 12 Commonwealth nations of the Caribbean have a combined population of just 6.18 million people (CIA, 2008 estimates) and include some of the smallest nations within the association's 53 countries and 2 billion persons. As in most island states, fisheries are an important component of the economy. Historically, coastal populations living below the poverty line have independently supplemented their food supply with resources from the sea.

The Commonwealth countries of the Caribbean include limestone islands such as Antigua and Barbuda, Bahamas and Barbados; and volcanic islands such as Dominica, Grenada, St Kitts and Nevis, St Lucia, St Vincent and the Grenadines. They also include islands with more complex origins, such as Jamaica and Trinidad and Tobago; and the two larger mainland territories of Belize, located in Central America, and Guyana, located on the north coast of South America.

These nations are located between 5°N and 24°N latitude. Except for Guyana, each has a coast on the largely self-contained Caribbean Sea. With the exception of Belize, the nations also have an east-facing coast on the west central Atlantic Ocean. Here, fresh water flowing northwards from the Orinoco and Amazon rivers in South America significantly decreases the sea surface salinity, which is lowest in July and August and highest in December, January and February (Kelly *et al.*, 2000). These estuarine conditions are carried along the island chain by the circulation of the North Brazil, Guiana and North Equatorial currents (Fratantoni and Glickson, 2002), taking with them nutrients that increase marine productivity and support the development of fisheries off the Guiana Shelf and northwards along the islands.

## Anticipated Climate Change in the Commonwealth Caribbean

A literature review reveals that while climate change is occurring in the Caribbean, the rate of change and even the direction of change remain unclear. Regional reviews are insufficient to make confident predictions (Cambers *et al.*, 2007) and, in the absence of empirical data, the authors have met in discussion groups and used expert opinion to predict likely outcomes based on the scenarios generated by Global Climate Change Models (GCMs) (Allison *et al.*, 2005). The varied environmental conditions and species composition of the fishery will result in a varied and localised, rather than a generalised, impact of climate change, and it will take more research to establish these patterns properly. Nevertheless, a growing body of evidence from specific sites supports the predictions based on the GCMs. One such analysis was conducted by the Intergovernmental Panel on Climate Change Working Group 1 (IPCC WG1) in 2007 (cited in Cambers *et al.*, 2007). A summary of these impacts is listed in Table 1 below:

**Table 1: Climate predictions for the Insular Caribbean**

| Climate parameter | Predicted change |
|---|---|
| *Direct Impacts* | |
| Air temperature | Increase of 1.8–4.0°C by 2099 |
| Hurricanes and storms precipitation | More intense with larger peak wind speeds and heavier precipitation |
| Precipitation | Unclear |
| Carbon dioxide | Reduction in pH of the oceans by 0.14–0.35 units by 2099 |
| Change in ocean current events – El Nino-Southern Oscillation (ENSO) | Forecasted increase in ENSO events |
| *Indirect Impacts* | |
| Global sea level | Rise of 0.18–0.59m by 2099 |
| Algal blooms and disease pathogens | Increase in number and impact |

Source: Adapted from Cambers *et al.* (2007) and Allison *et al.* (2005).

## Commonwealth Caribbean Fisheries

The figures vary widely but an aggregate of national reports to the Food and Agriculture Organization of the United Nations (FAO) counts some 56,000 persons that are employed directly in fishing and up to 100,000 more employed indirectly in its processing and distribution in the Commonwealth Caribbean (FAO, 2005). The fishery imported an approximate value of US$51m in fish and exported US$123.9m. This makes the region a net exporter of fishery products to the value of US$72.9m. See Appendix 1 for the values for individual countries. Note that most of the countries are net importers and that it is only Jamaica and the mainland territories of Guyana and Belize that are net exporters of fish products. These are the territories with a greater investment in commercial fisheries and they have a larger continental shelf on which to fish. This is not to say that the fisheries in the smaller territories are not important. They provide food for the local market and reduce some of the pressure to import fish products. They also provide some supplies to the local tourism and restaurant trade.

## The Structure of Fisheries in the Commonwealth Caribbean

Broad fishery categories are largely artificial as the fishers and the fish that they take are involved in an integrated environment with oceans to reef to shore to wetlands (Cavers *et al.*, 2002; IMA, 2006b; Cambers, 2007). Even the biological processes in the fish and benthos are integrated with early life history stages spent in the varying environments. In this scenario the fisheries should be seen as an integrated system with multi-gear being used as the fishers try to overcome seasonal variations in the life cycle of targeted species.

The industry can be categorised according to the level of investment in infrastructure in terms of both gear and landing facilities such that three groups emerge, namely: artisanal, industrial and recreational fishers. While the artisanal fishery was typified by a dugout canoe powered by oars, the investment here has grown to the point where, except for some Amerindian communities and the rural poor, the artisanal fishers have modernised and are using fibreglass reinforced boats and some improvement in the gear. The more *avant-garde* of these fishers are engaged in commercial fishing. They have adopted fish finders and GPS technology. These artisanal fisheries are based on near shore day trips in boats less than 10m long while industrial and semi-industrial vessels are greater than 10m long and undertake longer trips.

The fisheries have also been defined according to the physical environment that supports the industry including inland fisheries, coral reefs, coastal estuaries, and open ocean. In Trinidad and Tobago for example, the fishers engage in four major fisheries: coastal pelagic fishery, soft-bottom demersal (shrimp and groundfish) fishery, hard-bottom demersal fishery and oceanic (highly migratory) pelagic fishery.

## Reef fishery

Reef fisheries occur in the shallow-shelf reef, the bank-reef and the deep slope. Where there are substantial reefs, there is commercial, as well as, subsistence reef fishery. Smaller reefs can only support recreational and subsistence fisheries.

The shallow waters of the barrier reef and the three major atolls off Belize support the largest reef fisheries in the Commonwealth Caribbean. This barrier reef is the second largest in the world. It is nearly 260 km long. There are also 290 km of reefs around the outer atolls. The shallow lagoon between the coast and the barrier reef and those inside the coral atolls provide ideal conditions for the development of extensive seagrass beds. These provide forage for numerous commercial species including spiny lobster (*Panulirus argus*) and queen conch (*Strombus gigas*) (FAO, 2005). The fish species found in the reefs include hinds (*Serranidae*), parrotfishes (*Scaridae*), grunts (*Haemulidae*), surgeonfishes (*Acanthuridae*) and triggerfishes (*Balistidae*). These species are all considered overfished. The deep slope and bank reef fishery mainly targets snappers (*Lutjanidae*), the queen snapper (*Etelis oculatus*), silk snapper (*Lutjanus vivanus*) and vermillion snapper (*Rhomboplites aurorubens*).

Extensive reefs are found in Jamaica on the island shelf and the offshore Pedro and Morant Banks. The Bahamas reef fisheries occur on the shallow waters of the Little Bahama Bank and the Great Bahama Bank. Reef fisheries are also found on the shallow island shelves behind fringing reefs off Barbados, Antigua and Barbuda, St Lucia and St Kitts and Nevis and Tobago.

In Belize the landings for lobster averages about 278 MT. Average conch landings were

241 MT (FAO, 2005). All of the conch and lobster fisheries are considered fully exploited or over exploited. Stringent management quotas were adopted to encourage sustainable harvesting of these resources. The fishery comprises mainly of small (5–10 m) fishing vessels with outboard engines (15–75 hp) and simple capture methods. The methods include gill-netting, lobster and fish traps, casitas (lobster shades), trolling, hand-line fishing and free diving. Most fish potting occurs during the off-season for coastal pelagic fishery (June to December).

## Coastal fishery

Coastal fisheries include artisanal and industrial fishers who target specific demersal and pelagic species on a commercial scale. The largest of the commercial fisheries is the shrimp fishery. The artisanal vessels involved in coastal fishery employ simple gear such as 'a-la-vie', switchering, and trolling/towing lines, and multifilament or monofilament gillnets. The principal species targeted are carite (*Scomberomorus brasiliensis*), wahoo (*Acanthocybium solandri*), kingfish (*Scomberomorus cavalla*) and sharks.

Artisanal, semi-industrial and industrial trawlers are involved in the shrimp fishery. The shrimp fishery is important in Belize, Guyana and Trinidad. Trinidad shrimp fishery is shared with Venezuela in the Gulf of Paria. In 1998 there were 114 registered trawlers with 244 fishermen directly involved in this fishery (Soma, 2003). The Guyanese shrimp fleet consisted of 73 shrimp trawlers and 48 seabob/finfish vessels. There were five species of shrimp that are used for commercial purposes. These species include *Farfantepenaeus subtilis* (brown), *F. notialis* (pink), *P. brasiliensis* (pink spotted), *Lithopenaeus schmitti* (white), and *Xiphopeneaus kroyeri* (seabob, honey or jinga) (CRFM, 2004). The brown and pink shrimp were the most exploited of the five species (Ferreira and Soomai, 2001). In Belize the marine shrimp fishery landed 74 MT in 2004 with pink shrimp (*P. dourarum*) being the dominant catch. Four locally owned trawlers operate in this fishery (FAO, 2005).

## Ocean pelagic fishery

Oceanic fisheries are usually pelagic and are undertaken for both commercial and recreational purposes. The largest of the commercial oceanic pelagic fisheries is the flying fish and associated species fishery. Game fish tournaments attract attention for the larger pelagics. Many of these large pelagics are also caught to support the local restaurant trade.

The oceanic fishery involves commercial fishers who have invested in more expensive gear. Their fishing vessels include longliners, semi-industrial, multi-purpose commercial vessels and recreational vessels as well. The vessels are usually 14–23 m in length with diesel engines of about 180–350 hp and are equipped with echo sounders, fish finders, GPS, communication equipment and hydraulic equipment for setting and retrieving gear. There are fewer boats employed in this fishery compared to the artisanal fishery. The crew is usually about six persons compared to the three persons in an artisanal boat. Industrial fishers can operate further from the land and undertake longer fishing trips that last from seven to 15 days because of the communication and safety equipment, supplies and the refrigeration facilities to preserve the catch. They operate all year round with peak activity between November and April.

The principal species targeted in this fishery are the tunas: yellowfin tuna (*Thunnus albacares*), bigeye tuna (*T. obesus*), albacore (*T. alalunga*), blackfin tuna (*T. atlanticus*), swordfish (*Xiphias gladius*), and dolphin fish (*Coryphaena hippurus*). Sharks and carite are usually caught in the bycatch. Most of these species are highly migratory.

There is a catch quota system implemented by the International Commission for the Conservation of Atlantic Tunas and Tuna-like Species (ICCAT) and for Trinidad and Tobago it is 42 tonnes of swordfish per year. Trinidad and Tobago fishers usually reach this quota by March of each year (Lalla, 2002).

Barbados and Trinidad and Tobago share a substantial migratory pelagic fishery in the flying fish and associated species (Potts *et al.*, 2001; United Nations, 2008). Flying fish accounted for 70–90 per cent of the total weight of pelagic landings at beaches on the leeward side of Tobago in 1996 (Potts *et al.*, 1999). While the fleet in Barbados must venture further offshore in larger vessels, the smaller vessels in Tobago operate closer to shore. The Barbados flying fish fleet consisted of 250 vessels with a range of up to 50 km from the coast. They landed approximately 1582 MT of flying fish, 64 per cent of total annual landings (FAO, 2005). The fleet in Tobago consisted of 75 pirogues and one iceboat in the period 1988–1991. There are approximately 125 fishermen employed in the flying fish fishery in Tobago. They land approximately 1000 MT of flying fish annually (Potts *et al.*, 2001).

## Inland fisheries

Inland fisheries have not traditionally been of significance for commercial purposes and are pursued largely for subsistence and recreation. The fisheries occur in the ponds, mountain streams and coastal marshes such as the Black and Cabarita River morasses in Jamaica and the coastal marshes in Guyana. With the introduction of aquaculture production the value of the inland fisheries has been increasing. In 2004, the volume of total farmed shrimp exports from Belize were 16.86 million pounds valued at Bz$84.28 million (FAO, 2005). The Pacific white shrimp (*Litopenaeus vannamei*) produced on fourteen farms is the most important of the species cultivated. The commercial farming of tilapia is another of the significant aquaculture operations in Belize and Jamaica. In 2004 tilapia production in Belize reached 385 MT with fillet exports valued at Bz$1.1 million. Other small scale fish farming includes native finfish cichlids, such as the bay snook (*Pentenia splendida*), the crana (*Cichlasoma uropthalmus*) and the tuba (*Cichlasoma synspilum*) and the Nile tilapia (*Oreochromis niloticus*).

Jamaica has a substantial inland aquaculture programme geared at producing mainly tilapia for both the local and American markets. The production in 2002 was estimated at 6000 MT on farms located in the central plains of Clarendon and St Catherine. There are approximately 180 fish farmers. Tilapia accounts for 90 per cent of the food fish production (FAO, 2005).

## Impact of Climate Change on Commonwealth Caribbean Fisheries

The analysis that follows examines the impact of the projected climate change on the types of fisheries that have been described for the Commonwealth Caribbean.

### 1. Storms and severe weather events

Mangrove and seagrass beds are the nursery areas for many species of reef fish, conch and lobsters, which spend their early life history stages in these environments before returning to the reefs and ocean (Acosta *et al.*, 2007). The destruction of these related habitats would reduce the fish available for capture. Category 4 Hurricanes can severely damage mangroves as was seen in Hurricane Gilbert in Jamaica (Mahon, 2002). Bottom scouring of seagrass beds and other marine habitats that lie in the direct path of the storm result in replacement by algae, which reduce the habitat available for new coral colonies to attach (Fourqurean and Rutten, 2004).

The hurricanes cause differential mortality rates for each of the reef biotic communities (Magorien, 2004). These effects were expected to be greatest at shallow depths, where wave action is strongest. However shallow corals are thought to have adapted to wave action and hurricanes so that considerable damage is more likely to be found in deeper reefs where corals seldom experience wave action under normal conditions (Mahon, 2002).

Reefs in the Commonwealth Caribbean may already be shaped by the frequent storms such that the present reef structure consists of a larger proportion of massive corals that can more effectively withstand hurricanes than branching corals. Selective disturbance caused by hurricanes results in relative shifts in the plant and animal organisms in the reef to produce sustained balance in time (Magorien, 2004). The effects of hurricanes should therefore be considered in the long term. Hurricanes can directly reduce the physical complexity of coral reefs and the abundance of living corals. However, the increasing intensity of the storms and the frequency with which they strike could destabilise the reef ecology and lead to degradation as there is less time between storms during which the reef can return to equilibrium.

Marine community organisms are affected in the short term by the sudden decline in habitable space. Intense bottom scouring of the seagrass beds, beach overwash and turbidity in the coastal waters can re-suspend heavy metals. Increased dissolved phosphate levels for instance, can cause algal blooms, which would use up the biologically available oxygen and result in mortalities to the reef organisms (Magorien, 2004). In the hard bottom environments the corals are broken and transported to rubble heaps. Adult fish were observed to move to least damaged areas of the reef after hurricanes (Mahon, 2002).

Adult conchs bury themselves as storms approach and when sea conditions are rough this action reduces their availability to the fishers (Delgado *et al.*, 2004). However juvenile conchs were observed washed ashore in Antigua after Hurricane Luis (Mahon, 2002). This may have affected the availability of conch for recruitment in the subsequent fishery season. Peak larval settlement of lobsters takes place during the hurricane season therefore it is possible that advection of water masses by storms could affect the settlement of lobster larvae and subsequent recruitment. Following Hurricane Luis, Antigua had an increase in lobster catches, which could affect recruitment in subsequent harvest periods.

Of the human impacts, the flying fish industry in Tobago suffered extensive damage after Hurricane Flora in 1963. The Government supplied subsidies and credit facilities to resuscitate the industry (Potts *et al.*, 2001). Antigua and Barbuda suffered a loss of about 16 per cent of its fleet and a further 18 per cent was damaged during Hurricane Luis. Further losses were sustained after Hurricane Hugo. The Jamaican fishery has also experienced losses due to Hurricanes Gilbert and Mitch. Hurricane Ivan devastated the fishery equipment and supplies on the Pedro Banks and the South Shelf of Jamaica. The FAO estimated that six months is required for the full recovery in the landings of reef fisheries (Mahon, 2002).

Traditionally there is a windy season in the Commonwealth Caribbean fishing grounds that peaks from February to April (IMA, 2006a). Increasing extreme weather events in this season should cause further increases in costs by increasing travelling times to the fishing grounds, increasing fuel costs due to rough seas, increasing labour costs due to working in bad weather conditions, and increasing maintenance costs due to damage to the vessel, equipment and fishing gear. Destruction of fish traps during the windy season is one of the main costs caused by weather. During the windy season, and within the hurricane season from July to October, the fish move to deeper waters (Mahon, 2002). Catch is reduced. There is likely to be a reduction in fishing effort, which has the effect of increasing prices to consumers.

In general the increasing intensity of hurricanes and tropical storms will reduce the level of effort as most fishers operate in small ill-equipped boats that are not able to harvest in adverse weather. Those that venture out are exposed to increasing risk including loss of equipment, gear and of life. The fishers are restricted during the first two or three months of the year when high winds and rough seas can curtail operations for one to three week periods. The hurricane season also curtails fishing activity as the storms pass through the region during July to October. The net effect is an increase in the cost of landing the fish and because of the reduced supply, the price to the consumer is increased. Another school of thought suggests that the reduced fishing effort may lead to a recovery in the fish resources in the long term. However the possible effects of stock recovery may be negated by the increased fishing effort in the short term as displaced workers from other damaged sectors of the economy enter the open access fishery. This response was observed in Antigua and Barbuda after Hurricane Luis.

The personal risk to artisanal fishers is increased when there are storms and hurricanes. Usually these fishers do not carry safety and communications technology on their boats. Another impact on these fishers is seen in their communities as they may occupy low-lying sites along the coast. The storms may flood or destroy their homes as well as their landing facilities (IMA 2009). Aquaculture farms in Jamaica suffered 25 per cent loss of farm infrastructure and 62 per cent loss of fish stocks and equipment after Hurricane Ivan passed in 2004.

## 2. Increase in average sea surface temperatures and year-to-year climate variability

Storms can have damaging effects on the fishery in the reef resulting in a reduction in effort as well as destruction of the fish habitat. However the storms can also have a beneficial role. They can cause mixing of the surface waters to reduce the sea surface temperatures resulting in reduced bleaching of corals and less stress on pelagic organisms (Wilkinson and Souter, 2008).

The highest recorded sea surface temperatures in the Caribbean occurred in 2005, when sea surface temperatures were elevated above the summer maxima by 1 or 2 degrees for at least 4 months (Wilkinson and Souter, 2008). These sustained elevated temperatures were above the normal threshold and caused coral bleaching. Corals bleach when the coral polyp host becomes stressed through excessive production of toxic compounds by the symbiotic zooxanthellae (algae). The host reacts by releasing the symbiotic algae, causing the coral to appear white and very susceptible to disease. Many of the corals die, but under improved temperature situations they are known to recover. However, they tend to experience reduced growth and reproduction in the following season.

The 2005 bleaching in the Caribbean occurred almost simultaneously with major outbreaks of coral diseases (Wilkinson and Souter, 2008), resulting in reduction of live coral cover throughout the region and an increase in algal cover. Although many corals started recovery at the onset of winter with cooler seawater temperatures, coral diseases broke out and resulted in significant losses of coral cover, notably along the coast of Florida, in Belize, the Virgin Islands, and the Lesser Antilles. Bleached corals are said to be starving and lacking reserve lipids, which results in them being stressed and susceptible to disease.

The prolonged stress on corals due to the dangerously elevated sea surface temperatures resulted in the most severe coral bleaching and mortality ever recorded in the Caribbean (Wilkinson and Souter, 2008; personal communication Buccoo Reef Trust Directors). The French West Indies experienced 25–52 per cent coral mortality; Barbados recorded its most severe bleaching episode in history. In Tobago some 66–80 per cent of the coral cover was bleached. On average, the accumulated Caribbean thermal stress during the August–November 2005 period was greater than had been experienced by these reefs during the previous 20 years combined.

There can be a decline in fish density and biomass in coral reefs because of permanent coral cover reduction after several bleaching events. There is a lack of information on the response of tropical fish to temperature increases. However, the degradation of the fishery environment will undoubtedly reduce fish diversity and abundance and an increase in sea surface temperature is predicted to drive species ranges towards colder waters (FAO, 2008). This could result in extinctions where the dispersal capabilities are limited or suitable habitats are unavailable. Distribution and abundance may be reduced as the temperature increases cause changes in growth, survival or reproduction rates of the target species. This alteration in the distribution of fish will undoubtedly have an impact on the nature and value of commercial fisheries, but there are few hard facts available about the effect of the 2005 bleaching event on the fisheries in the Commonwealth Caribbean.

Wilkinson and Souter (2008) reported that some of the symptoms noticed when corals are damaged include particularly low fish populations, outbreaks of coral diseases and corals struggling to grow in poor quality, dirty waters or when smothered by algae. They advised that many of the reefs of the Caribbean were already stressed and in decline when the major climate change events of 2005 took place. They also pointed to the fact that islands and mainland countries of the Caribbean are highly dependent on coral reef resources. A World Resources Institute Reefs@ Risk analysis in 2000 estimated that the Caribbean coral reefs earned between US$3,100 million to $4,600 million each year from fisheries, dive tourism, and shoreline protection services; however 64 per cent of these same reefs were threatened by human activities, especially in the Eastern Caribbean, most of the Southern Caribbean, Greater Antilles, Florida Keys, Yucatan, and the nearshore parts of the Mesoamerican Barrier Reef System. All these areas suffered severe bleaching damage in 2005. It has been estimated that the coral loss could cost the region US$140 million to $420 million annually.

The need for the ocean pelagic vessels to travel longer distances to engage in fishing will increase the cost of the trip and increase the price of fish to the consumer. Additionally, there may be transboundary issues as migratory species are a shared resource that must be managed across international borders (McConney, 2000). There may be a need to re-evaluate the status of the stock and to grant more licenses for fishing in the territorial waters of other states. This is causing many of the states to regularize the boundaries of their Exclusive Economic Zone (EEZ). The arbitration between Barbados and Trinidad and Tobago led to the establishment of their maritime boundary and encouraged bilateral negotiations for sharing the flying fish and associated species resources (United Nations, 2008).

## 3. Increase in rainfall

Increased rainfall caused by climate change events should affect the nearshore environments and increase the extent of mangrove areas on the landward fringe. Growth rates and leaf litter in the mangrove should also increase. This would result in a more accommodating environment for the early life history stages of the reef fishery. However, the increased rainfall could also reduce salinity in the wetlands and nearshore.

The larger volume of mud, sediments and land-based pollutants transported by rivers after severe weather events would reach the shore, disrupt the ecology on the coast and reduce visibility in the sea. There should be no lasting impact of the sediment plume if it is quickly dissipated by the coastal currents. However the increasing frequency of large rainfall events and the runoff that follows may not allow for adequate removal of the suspended solids and nutrients, which could lead to algal blooms and eutrophication in localized areas. The increased rainfall should reduce fishing effort as the artisanal fishing vessels are not able to withstand the rains. Additionally, diving and snorkeling on the reef would be limited due to the sediment plumes and reduced visibility on the shallow reef.

On the other hand, decreased rainfall or periods of severe drought would lead to increased evaporation, resulting in reduced wetland habitats and limiting the proper growth and development of the early life history stages of reef fishes.

Coastal regions inshore can be affected by land run off and land based sources of pollution. While the rivers in the Commonwealth Caribbean are relatively small, the Orinoco and the Amazon provide substantial change to this environment and the effects of increased rainfall in these rivers should be considered. The northward dispersal of the sediments and other nutrients in these rivers affect the entire eastern Caribbean. Increased rainfall in these basins has been linked to major disruption in the fisheries in the Commonwealth Caribbean (Siung-Chang and Lum Kong, 2001). There can be an increase in pollutants in the waters. On the other hand, the increase in supply of nutrients should foster improved offshore marine production systems in the Caribbean. Inland fisheries could be affected as the fish habitat in the streams and ponds may be flushed at times when the fish are spawning.

## 4. Increase in carbon dioxide

Reefs are expected to be worst affected by the increase in carbonation and thus acidity in the waters. This has implications for the structure of the reef and for the calcifying organisms such as corals, molluscs, gastropods and sea urchins which make shells and plates from calcium carbonate (Wilkinson and Souter, 2008). The entire reef structure may collapse and this would certainly affect the fish abundance and diversity.

## 5. Change in ocean currents events – ENSO

Between November and May there are strong breezes and rough seas as a result of the offshore continental winds. Especially in the eastern Caribbean the sea is reported to surge intermittently. These surges stir up shallow seas down to depths of 20–30 m. The fish are reported to leave the affected areas and move to deeper waters. Trap catches are reduced. However, the surges cause lobsters to leave their hiding places and move about so they are more easily trapped (FAO, 2005). From June to October the opposite situation occurs, the sea is calm except for occasional storm events. Catches of fishes are high and those of lobster are low.

Conchs also follow a seasonal pattern. During the surges, they bury themselves in the sand and are difficult to catch. Deep reef fish are not observed to be affected by the surges. Changes in the seasonal cycle of these surges can affect the availability of these resources to fishers (Mahon, 2002).

There are likely to be small and medium scale changes in coastal circulation. This could impact on the coastal fishery as the planktonic early life history stages of some aquatic organisms and fish depend upon currents to transport them to the nursery areas. The change in strength and direction of winds and in turn the change in these nearshore currents could affect recruitment of fish.

Ocean currents are related to areas of upwelling. Changes to the upwelling in the

Guianas-Brazil shelf downstream of island passages and off Venezuela are known to influence fishery production and may be influenced by climate change. Many ocean pelagic species move within the circulation pathways of ocean currents. Changes in the current patterns could therefore affect the migration pathways of these fish resources.

## 6. Sea level rise

Sea level rise can have a positive effect on wetlands such as mangrove swamps, which may retreat shoreward, resulting in a gain of habitat for the early life history stages of the reef fishery, but in general, responses to sea level rise would be variable since there is a wide range of wetland types and geomorphic settings in the region (Cambers *et al.*, 2007). As an example of a negative impact, beach seining is carried out from the shore and if there is erosion resulting from sea level rise, the suitable beaches for seining may be eliminated. This type of fishery is carried out at the subsistence level and may affect the availability of local fish products.

Sea level rise may affect the low-lying landing sites and residences of the artisanal fishers. Increased erosion on the shore has made the landing of vessels very difficult (IMA, 2009). Fishers have to toil to remove their engines and nets from the boats. Winches have been installed but these need regular maintenance, adding cost to the fisheries. Many of the landing sites are located in the active tidal zone and require a further setback from the sites where erosion is taking place (Cambers *et al.*, 2007). The industrial fisheries will be particularly affected since few landing sites can accommodate their larger vessels and specialized facilities. The high cost of replacing the infrastructure can cause major disruption to the industry until the recovery takes place (UNFCCC, 2007). Salt-water intrusion as a result of sea level rise can negatively affect the habitat in the marshes and cause a decline in the stocks (FAO, 2008).

## 7. Increase in disease pathogens

The SEFSC (2002) reported that coral diseases are much more prevalent in the Caribbean than in the Pacific. At least ten coral diseases are present in the Caribbean. Since the gene pool in this small population in the Caribbean is smaller than in the Pacific this means that there is less genetic diversity and therefore resistance to pathogens and disease is lower than among larger coral populations in other regions.

There are increasing reports of new diseases and epidemics affecting corals; but how these diseases affect the fish and other benthos in the reef is unknown (Bouchon *et al.*, 2008). By 2007 staghorn and elkhorn corals (*Acropora cervicornis* and *A. palmata*) were experiencing major losses across the Caribbean after repeated outbreaks of white plague and other infectious diseases and bleaching of the corals (Wilkinson and Souter, 2008).

## 8. Increase in algal blooms

There are periodic algal blooms that impact fisheries. The resulting mass mortalities of fish can alter the marine food chains through adverse effects on eggs, juveniles and adult fish, turtles, seabirds and mammals (Luttenburg *et al.*, 2000). Harmful algal blooms are increasing worldwide in frequency, distribution and impact (National Assessment Synthesis Team, 2000). A fish kill that affected several countries in the south-eastern Caribbean was linked to increased water temperatures and the transport of a pathogen thought to be in the Orinoco Discharge (PAHO, 2000; Siung-Chang and Lum Kong, 2001).

### Conclusion

Fisheries can expect both positive and negative impacts as a result of climate change in the Commonwealth Caribbean. Interactions between and amongst the environmental systems may add to or diminish these effects. The dearth of information on climate change in the Caribbean precludes any definitive statement on these effects, and this uncertainty is compounded by changes in the food chain that are directly induced by human activity, such as overexploitation of the predatory and grazer species. There are also episodes of land-based and other pollutants being released into the waters through sewage, effluent from industry and ballast water.

The fishers are themselves preoccupied with issues such as loss of use of traditional landing sites and fishing grounds to other fast growing industries, as expressed by Claxton Bay fishers, Trinidad (IMA, 2009). There are problems with marketing and distribution. Processing facilities are at a minimum. There is an increasing exposure to piracy, both on land and at sea in an atmosphere of rising levels of crime and the landing of illicit and contraband items. These and other problems cause the stakeholders in the fishing industry to be preoccupied to the point where the impacts of climate change are not receiving the attention they deserve. It is certain that climate change has already begun to occur and will have impacts on the fishery. Trends must be identified and steps taken to adapt to the expected changes.

## References

Acosta, A., Bartels, C., Colvocoresses, J., and Greenwood, M.F.D. (2007) Fish assemblages in seagrass habitats of the Florida Keys, Florida: Spatial and temporal characteristics. *Bulletin of Marine Science*, 81 (1), pp. 1–19.

Allison, E.H., Adger, W.N., Badjeck, M.C., Brown, K., Conway, D., Dulvy, N.K., Halls, A., Perry, A., and Reynolds, J.D. (2005) *Effects of climate change on the sustainability of capture and enhancement fisheries important to the poor: analysis of the vulnerability and adaptability of fisherfolk living in poverty*, Project No. R4778J Final Report. Fisheries Management Science Programme (FMSP) Department (DFID).

Bouchon, C., Portillo, P., Bouchon-Navaro, Y., Max, L., Hoetjes, P., Brathwaite, A., Roach, R., Oxenford, H., O'Farrell, S., and Day, O. (2008) Status of Coral Reefs of the Lesser Antilles after the 2005 Coral Bleaching Event Ch. 8, pp. 8–103 in *Status of Caribbean Reefs after bleaching and Hurricanes in 2005*. Clive Wilkinson and David Souter (Eds).

Cambers, G., Claro, R., Juman R., and Scott, S. (2007) *Working Group Report, Climate Change Impacts on Coastal and Marine Biodiversity. University of Puerto Rico Sea Grant College Program, Puerto Rico* (Institute of Oceanology, Cuba/Institute of Marine Affairs, Trinidad and Tobago/Fisheries Department, St Lucia).

Caribbean Regional Fisheries Mechanism (CRFM) (2004). *Report on the Shrimp and Groundfish Workshops conducted in Guyana (2003), Belize (2003) and Trinidad* (Belize City, Belize) pp. 1–93.

Cavers, C.V., Furley, P.A. and Mumby, P.J. (2002) *An Assessment of the relative importance of a mangrove island within a Caribbean atoll in the early life history of coral reef fishes.* Poster presentation. Geography, School of Earth, Environmental and Geographical Sciences, University of Edinburgh, U.K.

Central Intelligence Agency (CIA) (2009) *World Fact Book – Commonwealth Caribbean Countries* (individually).

Delgado, G. A., Bartels, C. T., Glazer, R. A., Brown-Peterson, N. J., McCarthy, K. J. (2004). Translocation as a strategy to rehabilitate the queen conch (Strombus gigas) population in the Florida Keys. *Fisheries Bulletin* 102 (2) pp. 278–288.

FAO (2005) *Country Profile: Structure and Characteristics of the Fishing Industry* (individual countries).

FAO (2008) Report of the FAO Expert Workshop on climate Change Implications for Fisheries and Aquaculture. *FAO Fisheries Report* No. 870.

Ferreira, L. and Soomai, S. (2001) Trinidad and Tobago Shrimp and Groundfish Fisheries. Regional reviews and national management reports – Fourth workshop on the Assessment and Management of Shrimp and Groundfish Fisheries on the Brazil-Guianas Shelf Cumaná, Venezuela, 2 to 13 October 2000. FAO/FISHCODE Project (GCP/INT/648/NOR). Fisheries and Aquaculture Department, FAO and CARICOM Fisheries Resource Assessment and Management Program (CFRAMP).

Fourqurean, J., and Rutten, L.M. (2004) The Impact of Hurricane Georges on soft bottom back reef communities: site and species specific effects in south Florida seagrass beds. *Bulletin of Marine Science*, 75 (2) pp. 239–257.

Fratantoni, D.M. and Glickson, D.A. (2002) North Brazil current Ring Generation and Evolution Observed with SeaWiFS. *Journal of Physical Oceanography*, American Meteorological Society. 32. pp. 1058–1074.

Heron, S., Morgan, J., Eskin, M. and Skirving, W. (2008) Hurricanes and their effects on Coral Reefs, Chapt. 3, pp. 31–36 in C. Wilkinson and D. Souter (Eds) *Status of Caribbean Reefs after bleaching and Hurricanes in 2005.*

IMA (Institute of Marine Affairs), (2006a) *State of the Marine Environment Report for Trinidad and Tobago.* Unpublished.

IMA, (2006b) *Assessment of Kilgwyn Swamp.* Prepared for the Tobago House of Assembly.

IMA (2009) *Needs Assessment of the Stakeholders in the Fisheries Sector.* Prepared for the Seafood Development Company, Trinidad and Tobago.

Kelly, P.S. Lwiza, K.M.M. and Cowen, R.K., Goni, G.J. (2000) *Low-salinity pools at Barbados, West Indies: Their origin, frequency and variability* (Marine Science Research Center, State University of New York at Stony Brook and Atlantic oceanographic and Meteorological Laboratory, National Oceanic and Atmospheric Administration, Miami, Florida. American Geophysical Union).

Lalla, H. (2002) National Report of Trinidad and Tobago, Fisheries Division, Ministry of Food Production and Marine Resources, St Clair Circle St Clair, Port of Spain, Trinidad, W. I.,

Magorien, R. (2004) *The Atlantic Hurricane and its Effects on Caribbean Coastal Systems.*

Mahon, R. (2002) *Adaptation of Fisheries and Fishing Communities to the Impacts of Climate Change in the CARICOM Region*, Issues paper, Prepared for the CARICOM Fisheries Unit, Belize City,

Belize, as input to the planning process for the project Mainstreaming Adaptation to Climate Change (MACC) of the Caribbean Centre for Climate Change (CCCC). (Washington: Organization of American States).

McConney, P. (2000) *Multi-objective Management of Inshore Fisheries in Barbados: A Biodiversity Perspective*. Coastal and Marine Management Program, Caribbean Conservation Association.

National Assessment Synthesis Team (2000) *Climate Change Impacts on the United States The Potential Consequences of Climate Variability and Change Overview: Coastal Areas and Marine Resources* (US Global Change Research Program).

Pan American Health Organization (PAHO) (2000) Fish Mortality in southeastern Caribbean countries. *Epidemiological Bulletin* 21 (2) pp. 8–10.

Potts, A.C., Thomas, A.D., Nichols, E., Lovelace B. and Henry L. (1999) The flyingfish industry of the Republic of Trinidad and Tobago, pp. 27–39 in: *FAO Fisheries Report* no. 613 SLAR/R613. Report of the first meeting of the WECAF AD HOC Flyingfish working group of the eastern Caribbean, Bridgetown, Barbados, 22–24 September 1999. FAO, Rome, Italy.

Potts, A.C., Thomas, A.D. and Nichols, E. (2001) *An economic and social assessment of the flyingfish (pelagic) fishery of Tobago, Trinidad and Tobago*. Department of Marine Resources and Fisheries; Scarborough (Trinidad and Tobago). 15 pp. Dec 2000. II Meeting of the WECAFC *ad hoc* Flyingfish Working Group of the Eastern Caribbean. Cave Hill (Barbados); 8 Jan 2001.

Siung-Chang, A., and Lum Kong, A. (2001). Possible links between reef-fish mortalities in the south eastern Caribbean and South American river discharge (July–October 1999). *Bulletin of Marine Science*, 68, pp. 343–349.

Soma, K. (2003) How to involve Stakeholders in Fisheries Management – a Country Case study in Trinidad and Tobago, *Marine Policy* #27, pp. 47–58.

South East Fisheries and Science Centre (SEFSC) 2002. Coral Reef Research Plan, FY 2001 – FY 2005. National Marine and Fisheries Service (NMFS).

United Nations Framework Convention on Climate Change (UNFCCC) (2007) *Climate Change: Impacts, Vulnerabilities and Adaptation in Developing Countries*, Ch. 4, pp. 16–24.

United Nations, 2008. Arbitration between Barbados and the Republic of Trinidad and Tobago, relating to the delimitation of the exclusive economic zone and the continental shelf between them, decision of 11 April 2006 in *Reports of the International Arbitral Awards* XXVII pp.147–251

Wilkinson, C. and Souter, D., 2008. *Status of Caribbean Reefs after bleaching and Hurricanes in 2005.*

# Appendix I

## Fishery Information for Commonwealth Caribbean

| | Barbados | Trinidad and Tobago | Jamaica | St Lucia | Guyana | Dominica | Belize | St Kitts/Nevis | St Vincent and the Grenadines | Grenada | REGION |
|---|---|---|---|---|---|---|---|---|---|---|---|
| **Fisheries data** | 2002 | 2003 | 2002 | 2006 | 2004 | 2000 | 2003 | 2004 | 1999 | 2005 | TOTAL |
| Estimated employment | | 1.18% | | | | | | | | | |
| Primary sector | 2,000 | 5,100 | 20,000 | 2,319 | 6,500 | 2,843 | 1,672 | 600 | 2,500 | 2,400 | 45,934 |
| Secondary sector | 825 | 1,985 | | 120 | 6,000 | 60 | 123 | | 500 | 400 | 10,013 |
| Gross value of fisheries output: (US$m) | n.a. | 21 | 20.7 | 7.4 | n.a. | 13 | 53.4 | n.a. | 14.2 | 8.631 | 125.331 |
| **Trade** | 2003 | 2004 | 2002 | | | | | 2004 | | 2005 | |
| Value of imports (US$ m) | 13.5 | 14.275 | 8.3 | 6 | 1.07 | 3.1 | 2.574 | 2.227 | 1.149 | 2.465 | 51.96 |
| Value of exports (US$ m) | 0.869 | 9.388 | 39.7 | 0 | 54.2 | 0 | 16.239 | 0.00016 | 0.961 | 3.413 | 123.9012 |

Source: FAO, 2005, 2008.

Note: No FAO reports were available for Antigua and Barbuda, and The Bahamas. Also the reporting years vary as individual countries submitted reports at different times.

# 13 Marine Protected Areas as a Strategy for Sustainability

Marta Lang and Rowan Trebilco

## Introduction

This chapter describes the key role marine protected areas (MPAs) must play to restore and sustain fisheries in the Commonwealth and worldwide. We summarise current global MPA coverage, and how MPAs fit into international systems of law and policy including the frameworks of marine spatial planning and ecosystem-based management. The roles different types of MPAs can play in restoring depleted stocks, enhancing productivity by protecting key habitats, and contributing to the resilience of ecosystems in the face of environmental change and pollution are discussed. Social considerations and benefits, and important aspects of design and implementation are also considered. We illustrate key points with case studies from Australia, St Lucia and Kenya.

## 1. Improved management practices

Marine systems are critically degraded globally (Sala and Knowlton, 2006). They are threatened by a suite of stressors; most notably over-harvesting, pollution and the effects of climate change (Hughes *et al.*, 2005). 70–80 per cent of world fisheries are fully exploited, over-exploited or depleted relative to maximum sustainable yield (MSY) (FAO, 2009). 82 million tonnes of fish were removed from marine fisheries in 2006 (FAO, 2009), and global bycatch estimates range from 17.9–39.5 million tonnes per annum (Day, 2006).

Populations of large, valuable predatory fish have been estimated for some time to be at 10 per cent of pre-industrial levels in continental shelf waters and the open oceans (Myers and Worm, 2003). Having fished out large valuable fish, humans are now 'fishing down the food web' (Pauly *et al.*, 1998), and most natural refugia have been eliminated (Agardy *et al.*, 2003). The latest analysis now suggests most fish populations are at 10 per cent of their biomass before fishing began (Lotze and Worm, 2009).

The scale and severity of fisheries impacts has led to recognition that marine ecosystems are in urgent need of improved conservation and management practices; practices that ensure fisheries remain able to supply seafood and the other marine ecosystem services on which we rely, and that we avoid ocean life extinctions comparable to those occurring on land (Roberts and Hawkins 1999). It is clear that we need to be much more discerning about where and how we fish, and considerable commitment will be required to restore depleted stocks and degraded ecosystems (Jackson *et al.*, 2001).

### 2. International law and policy

In 1994 when the Law of the Sea Convention 1982 came into force, the assignment to coastal nations of full territorial management rights over seas out to twelve nautical miles offshore, was complemented by an Exclusive Economic Zone (EEZ) management jurisdiction to 200 nautical miles. The EEZ management jurisdiction came with obligations both to conserve and manage EEZ resources; and to regulate EEZ take at (or below) maximum sustainable yield (MSY). Fisheries within 12 nautical miles and in the EEZ must also be managed to preserve and protect the marine environment, and consistent with state party requirements set out in the Convention on Biological Diversity.

Until recently, emphasis has typically been given to managing fisheries at MSY, both within and outside the 12 mile limit, rather than on other objectives. However, it is now accepted that 'traditional' management methods such as MSY targets are inadequate for addressing the multiple anthropogenic impacts affecting marine ecosystems, which encompass over-exploitation of both target and bycatch species, habitat degradation via destructive fishing methods, pollution, coastal development, and other human activities (Halpern, 2003; Halpern *et al.*, 2008). With the depletion of most stocks below the biomass that supports MSY there is impetus for EEZ fisheries management to place more emphasis on implementing fisheries conservation measures, together with preservation of the marine environment. The tide is necessarily turning towards protection and restoration in fisheries.

### 3. A role for MPAs

At the World Summit on Sustainable Development in 2002, it was agreed that national networks of marine protected areas would be established by 2012 and overexploited fisheries rebuilt by 2015 (United Nations, 2002). This is a clear signal of international marine policy direction. The scientific advisory body to the Convention on Biological Diversity recommended in 2003 that marine reserves closed to all extractive use should form the core of national MPA networks, and the World Parks Congress of 2003 recommended that 20–30 per cent of every habitat in the sea should be given full protection from fishing (Roberts *et al.*, 2005). Meeting this target has a projected cost of between US$5 billion and $19 billion annually (Balmford *et al.*, 2004). This figure does not, however, consider gains from improved fisheries and ecosystem services, which are anticipated to greatly outweigh the costs (Balmford *et al.*, 2004; Costanza *et al.*, 1997)

## MPA Types, Coverage and Establishment

### 1. What are MPAs and how do they fit in?

Place-based approaches to management (often referred to as marine spatial planning) divide marine ecosystems into distinct management areas using available biophysical, ecological, political and socio-economic information, with management rules based on the values the area contains. MPAs, "areas permanently protected against at least one preventable threat" (Norse and Crowder, 2005), form a central component of marine spatial planning. Their location and degree of protection depends on whether areas contain (or typically used to contain) values we wish to safeguard or restore. MPA planning fosters and supports multiple species and ecosystem-level thinking.

## 2. Types of MPAs

While MPAs typically aim to limit extractive resource uses (most notably fishing), other non-extractive activities such as boat traffic, dumping of waste and tourism may also be regulated. MPAs encompass a potentially broad range of activity restrictions. The most restrictive category, 'no-take MPAs' or 'marine reserves', are 'places that are permanently protected from all preventable anthropogenic threats' (Norse and Crowder, 2005). Even within the marine reserve category, the level of restriction may vary from strict scientific reserves with access restricted to scientific research (often requiring difficult to obtain permits), to reserves that are closed to fishing but open to all non-extractive recreational activities. Generally speaking, marine reserves protect areas from fishing and habitat degradation, enabling resident populations to move back towards unexploited size structures and levels of productivity. Habitat structures are spared from destructive extraction methods and this benefits both resident and transient populations.

Less restrictive MPAs may be closed only to specific fisheries, or only at specific times of the year. Such 'time-area closures' may be employed to rebuild overfished populations, to protect seasonal spawning grounds, or to prevent interactions with valued non-target species (Norse and Crowder, 2005) such as sea birds and marine mammals around breeding colonies (Trebilco *et al.*, 2008).

## 3. Coverage

Major challenges remain to meet the goals for marine ecosystem protection identified by the scientific community, to translate ecosystem-orientated policy and scientific principles into MPAs in the water, and to work towards establishing networks (Day, 2006). The global MPA count increased from 1306 MPAs in 1995 to 5045 in 2008 (Spalding *et al.*, 2008). Yet less than 1 per cent of the ocean is protected – an order of magnitude less than the 20–30 per cent goal.

As demonstrated by Rodrigues *et al.* (2004) and Spalding *et al.* (2008), certain depths and ecosystems are more protected than others. Only 4.09 per cent of the global continental shelf area is incorporated within MPAs, compared to 12.1 per cent of the narrow coastal belt. Moreover, at one end of the spectrum 8 per cent of marine ecoregions have over 30 per cent of their area protected in MPAs, yet approximately half of all ecoregion types have less than 1 per cent of their areas protected (measured across the whole shelf). Protection is greatest in the tropical realms; temperate realms remain poorly protected. Given that that many MPAs lack effective management, these coverage estimates are an optimistic measure of the scope of habitat protection worldwide.

Publications in the peer-reviewed literature debate what proportion of the sea should be protected in MPAs. Recommendations depend in part on the goal of protection: stock viability, fisheries sustainability, ecosystem resilience, or the protection of every kind of marine community. Most studies predict maximum benefits with 20–40 per cent of fishing grounds closed (Roberts, 2007) – with at least 20 per cent of each habitat type in no-take reserves to safeguard fishery production and biodiversity (Roberts *et al.*, 2003).

### 4. Establishment approaches

The approach taken to MPA planning can have important implications for MPA performance in contributing to sustainable fisheries. Broadly, design approaches may be categorised as either *ad hoc* or systematic.

Under *ad hoc* approaches, reserves are essentially established opportunistically, and their location is generally driven by vocal leaders from within concerned communities or funding interests. Areas are typically chosen on human scales for management. Oceanographic zones for protection may be broadly categorised into coastal, offshore and deep sea, and the focus of *ad hoc* approaches is different for each:

- Coastal: participatory approaches for buy-in (and information feed in) to support ownership and enforcement.
- Offshore: managed closures to fishing methods, commercial fishing, or industrial factory ships. This is generally top down designation, supported by vessel monitoring.
- Deep sea: protection for key features including underwater islands of life such as seamounts, and convergence zones. These are implemented through regional agreements, spurred by top down pressure from annual United Nations oceans resolutions.

For each of these zones there are different associated knowledge inputs regarding the ecological system, enforcement challenges, and stakeholder groups – generally with increased corporate and foreign state interest further from shore.

In contrast to *ad hoc* approaches, systematic approaches to MPA planning feed available data and planning goals into systematic decision tools (typically computer based). The decision tools generate options for placement of protected areas that maximise performance and minimise conflicts on the basis of the identified planning goals (Sarkar *et al.*, 2006). The prescriptions of the decision tool are then typically modified according to political and social influences (Margules and Pressey, 2000). Ideally then performance is modelled, and re-iterated to achieve the best compromise between predicted ecological performance, and stakeholder, political and economic interests.

## An Ecosystem Management Orientation

### 1. Ecosystem-based management

The favoured approach to achieve the goals of ecosystem protection and restoration, in combination with maintenance and restoration of fisheries yields and other essential ecosystem services, is ecosystem-based management (EBM). EBM represents a shift in focus from individual species to ecosystems, taking into account the interactions and interdependencies between co-occurring species and habitats, as well as the linkages between habitats and populations across seascapes. Species-based ecosystem modelling becomes unfeasibly complex and lacks robustness once it contains more than a few species (Crowder *et al.*, 2008; Young *et al.*, 2007), and it is extremely difficult to account for all the complex interactions with the biophysical environment, in addition to humans. Therefore, place-based approaches are considered the most practical and feasible way to achieve EBM.

Key steps for applying EBM are to identify management objectives for the ecosystem, including natural and human goals, and to ensure that the governance structure matches the scale of ecosystem management (Ruckelshaus *et al.*, 2008).

## 2. Complementary management methods and effort restrictions

MPAs form a central component of place-based approaches to EBM, representing those areas deemed to require protection from one or more threatening processes. However, MPAs cannot achieve sustainable fisheries management alone, and represent only one component of EBM. There is a parallel need for careful consideration of management targets and approaches across all areas, and often for reduced fishing effort and carefully considered gear and take restrictions in fished areas.

Fundamentally, a mechanism for controlling exploitation methods, through complementary, layered management regimes (including MPAs) is required. These may include gear, take and effort restrictions, within specific zones where marine spatial planning is initiated or throughout an ocean jurisdiction. Highly mobile and long-lived species require particular additional protection within EEZs and territorial seas due to their inherent vulnerability to overfishing.

With this in mind, in addition to implementing MPAs, coastal states need to implement regulatory change in their territorial seas and EEZs to:

- Replace quotas – which only count fish landed legally, and ignore at sea undersized and quality discards – with effort restrictions on location, time and gear use, supported by vessel monitoring systems;
- Require fishers to keep what they catch, so they fish more selectively (Roberts, 2007);
- Regulate specific gear types – creating incentives that encourage the use of selective gear that minimises bycatch, and banning destructive fishing gear like trawls on valued hard habitats (Roberts, 2007);
- Protect demographically or ecologically important size-classes (such as reproductive females and the largest individuals) by prohibiting take of individuals above or below specified size bands (Smith *et al.*, 1998).

## 3. Stopping destructive fishing methods

Traditional fisheries management assumes there is enough quality habitat to support the growth of target species (Roberts *et al.*, 2005). This is a flawed assumption, as fishing gears often cause physical habitat degradation, or change the balance of predators and prey within a system such that habitat-forming species deteriorate (Roberts, 2007). Perhaps the most extreme example of a small-scale destructive fishing method is dynamite fishing, which has been outlawed in many places. However, many other fishing methods that still remain legal cause massive damage to marine habitats. Trawling effectively 'bulldozes' the bottom, destroying complex communities of sponges and corals that may have acted as nursery habitat for target and non-target species (Roberts, 2007). Areas damaged by trawlers may be extremely slow to recover, especially coral and slow growing deep-sea habitats like seamounts (Watling and Norse, 1998). Additionally, it is becoming increasingly evident that many fishing methods cause more cryptic damage not evident from material brought to the surface.

If fishing is a major driver of damage, preventing the use of destructive methods in MPAs can prevent further habitat loss. Such bans can also give damaged habitats an opportunity to recover, if they have not already been irreversibly damaged. 10 per cent of the world's coral reefs are estimated to have been damaged beyond recovery (Roberts, 2007).

## 4. Protecting key habitats

Protection of key habitats like the spawning and nursery grounds of mobile species (e.g. mangroves for many reef fishes) may involve MPA establishment some distance from the main fishing grounds. Feeding and breeding areas used by fish at specific life stages or times of year, and migration routes, can be identified for the most valuable large migratory fish (such as tuna) from literature (Hooker and Gerber, 2004). Aggregation points along migration pathways, such as seamounts, act as key habitat. Increasingly, research shows that 'homing' (ie returning home regularly) is widespread among fishes and vertebrates, including sharks returning to nursing grounds (Edren and Gruber, 2005); and cod to spawning sites (Robichaud and Rose, 2004). In such situations, protection of relatively limited areas may have great benefits for fisheries (Stefansson and Rosenberg, 2006).

Habitats used will differ between species, but the overlapping intersections of important habitat areas, used by multiple species, are obvious targets for MPA protection. Some deep sea features such as seamounts, and coastal features such as estuaries and mangroves are well-known habitat areas for many species. For some species, key habitat areas may not overlap with those of others, hence the need for supplementary single species measures.

## 5. Spatial planning, MPA zoning and multi-use areas

Key thinkers advocate for management systems where "no-take MPAs are strategically linked with, or an integral part of, multiple use protected areas" (Agardy *et al.*, 2003). In such 'multi-use' areas, "use is controlled to ensure long term conservation goals are not compromised" (Day, 2006).

Marine spatial planning is the proactive technique used to designate the places within multi-use areas that will be managed as MPAs, and the type of MPAs that will be implemented. Varied values within large MPAs may be addressed by a zoning approach, by which the MPA is sub-divided into management zones with differing levels of protection. The marine spatial planning process may also consider what controls are appropriate in areas that are fished. This approach was pioneered for the Great Barrier Reef Marine Park in Australia, where it was very successful (Fernandes *et al.*, 2005) (refer to Case Study 1).

## Case Study 1: Great Barrier Reef, Australia

The Great Barrier Reef Marine Park (GBRMP) encompasses 345,000 square kilometres off the Australian coast (an area greater than the United Kingdom, the Netherlands and Switzerland combined). It has had a zoning system since the 1980s, with a major revision implemented in 2004. A multiple-use zoning approach provides high levels of protection for specific areas whilst allowing reasonable uses, including commercial fishing, to continue in other zones. Local stakeholders including recreational and commercial fishers were actively involved in the planning process. In addition to zoning, other management 'tools', including permits, education, and more recently management plans, have been applied to regulate access and to control and mitigate impacts associated with human use of the GBRMP.

A systematic planning approach was adopted for the 2004 zoning revision, with specific protection goals including at least 20 per cent protection per 'bioregion', minimum levels of protection for all known habitats and special or unique features, and minimum sizes for no-take areas. More than 33 per cent of the Great Barrier Reef Marine Park is now in no-take areas (previously 4.5 per cent).

Each year about 24,000 tonnes of seafood (valued at around $120 million AUD) is harvested commercially in the GBRMP, and recreational (including charter) fishing continues to be a major use.

Sources: Fernandes *et al.* (2005); GBRMPA (2003).

## Ecological Considerations and Barriers

Area protection strategies assume populations and ecosystems will recover once fishing pressure is removed, but some may have been irreversibly damaged by exploitation (Pinnegar and Engelhard, 2008). While there is great potential for protection to contribute directly and indirectly to the restoration of depleted stocks, there have been several prominent cases where stocks fished to the point of collapse have not recovered once fishing pressure has been removed – a famous case is the Canadian Northern Cod fishery. Few examples exist of successful recovery of heavily exploited fisheries (Crowder *et al.*, 2008), but one exception may be Georges Bank where there was a large closure to mobile fishing gear, giving the intensively trawled bottom a chance of recovering (Roberts, 2007).

Recovery is difficult where ecosystem-scale shifts occur, and any long term species return relies on there still being a prey base, no other species take over of niche (Crowder *et al.*, 2008), sufficient intact habitat and recruitment back to the area at sufficient density to reproduce (Roberts, 2007). In addition, many species exhibit site fidelity and their stocks may take a long time to recover if wiped out in key habitat areas. Recovery may also be inhibited by incidental catch in commercial fisheries targeting other species, or bycatch by recreational fishermen.

Fishing preferentially targets larger fish species and larger individuals. This leaves populations of reduced average sizes and reproductive rates since big animals are more fecund than small (Roberts, 2007). Systemic take of fish above size at maturity is especially problematic. In a meta-analysis of fish recovery, Hutchings (2000) concludes recovery is linked to age at maturity, reproductive rate, fishing gear selectivity, and any ecosystem-level consequences of exploitation, for example, food webs altered by changes to species community structure. Large predatory fish at the top of the food chain are the most valuable in fishery terms. Evidence for recovery of long-lived, slowly maturing top predators is limited (Crowder *et al.*, 2008). Small fish species tend to reproduce faster, so cope better with periodic fishing and bounce back faster when fishing is suspended (Roberts, 2007). However short life histories may make species more susceptible to deleterious evolutionary effects of fishing (Law, 2007), which may also inhibit recovery.

## Socio-economic Considerations and Barriers

### 1. Closure controversy

Despite a long history of permanent area closures in some cultures, and recent use as a tool in Western fisheries management (Nowlis and Friedlander, 2005), MPAs (and marine reserves in particular) can be socially controversial. This is because, fundamentally, by protecting species and habitats MPAs restrict extractive resource user activities or displace them to non-protected areas. This can cause social tension and even conflict if imposed without consultation and engagement of users in the design process, as well as full communication of the anticipated benefits to the resource and community (Agardy *et al.*, 2003). Social inertia means the majority of MPAs are initiated when resources are already at dangerously low levels and there is 'little to lose'.

### 2. Social benefits

Area closures can offer significant social fisheries management benefits, including enhanced resource flows and social cohesion, particularly if part of a multiple use zoning approach that has been collaboratively and carefully designed. Concepts such as taboos and stewardship often have a history and resonance with local people (Bartlett *et al.*, 2009; Cinner and Aswani, 2007). If set up with consideration of such factors, MPAs can re-establish traditional links, pride and ownership in an area. This 'heart' link can mean small scale users become advocates for more sustainable systems of management. MPAs can be designed and implemented locally, and with a relatively simple rule framework – both of which enhance user compliance (Bartlett *et al.*, 2009).

Where coastal communities are empowered to come up with their own answers to overfishing through participatory processes overseen by trained facilitators over an extended period, ownership and self-enforcement can evolve. Establishing management structures such as local committees to work on area goals and management arrangements fosters ownership and cohesion and provides both a forum for agreement and a mechanism for delivery following agreement.

95 per cent of the world's fishers are from developing countries, catching 58 out of the 98 million tonnes of all fish harvested in 1999 (Berkes *et al.*). While industrialised vessels account for a large proportion of overall catches, small-scale fisheries employ 50 out of the 51 million fishers worldwide (Berkes *et al.*, 2001) and these fishermen take significant quantities of fish to feed their families and maintain livelihoods. By sustaining stocks into the future, MPAs offer economic and social benefits to fishers. White and Kendall (2007) conclude that "… reserve-based policies generated profits approximately equal to or greater than those attainable under optimal conventional management".

An MPA itself presents a new potential employment source in areas with tourism, operating as an alternative or supplementary income source to fishing (Russ and Alcala, 1998). A proportion of fees can be injected back into management.

## 3. Funding and monitoring

The development of community-level funding streams can critically enhance the chance of management regimes surviving across the time scales required for MPAs to enable recovery of exploited populations and degraded habitats. Enforcement is crucial to the effectiveness of MPAs, and monitoring is important in assessing whether MPAs are meeting management goals. However, both enforcement and monitoring can be prohibitively expensive. Involving locals in enforcement and monitoring can help reduce costs, and reinforce the value of enforcement efforts. However, recovery can be slow and results may not be observable in the early years following establishment. Ongoing mentoring and follow-up by facilitators can reinject critical enthusiasm during this time.

## 4. Governance and enforcement in the offshore

The potential to involve local stewards in the implementation, enforcement and monitoring of MPAs is lacking for offshore habitats, as locals generally only fish waters close to shore. Offshore MPA policing (while do-able remotely via increasingly widespread vessel monitoring systems) is expensive and non-existent in many countries. Unlike fishers living in coastal communities, parties involved in offshore fishing ventures are unlikely to have a generational connection to the fishery. Further, there are huge financial investments in industrial fishing, and fishers have strong incentives not to share knowledge about species locations (Wilson, 2006). Also, historical freedoms can feed into present day resistance to activity restrictions by users (Jones, 2001; Westcott, 2006).

## The Role of MPAs in Restoring Sustainability to Fisheries

### 1. Design strategies

The optimal design and placement of MPAs is extremely complex, and a broad body of literature is developing on the topic (Almany *et al.*, 2009; Baskett *et al.*, 2005; Botsford *et al.*, 2003; Carr *et al.*, 2003; Dayton *et al.*, 2000; Fernandes *et al.*, 2005; Halpern *et al.*, 2006; Lundquist and Granek, 2005; Nowlis and Friedlander, 2005; Palumbi *et al.*, 2003; Roberts *et al.*, 2003; Stefansson and Rosenberg, 2006; Wood and Dragicevic, 2007).

In general, the scale of closures should be the same order of magnitude as the movement range of the species you aim to protect (Gell and Roberts, 2003). If a goal is to sustain all species along, say, a coastline, the area protected should be designed for the species with the largest range. For coral reefs, even MPAs less than one km in diameter can enhance local fisheries, especially if networked with others (Gell and Roberts, 2003). For mobile estuarine species, MPAs in the order of tens of square kilometres have enhanced recreational fisheries in Florida. MPAs ten thousands of square kilometres have restored whole fisheries previously in decline – such as Georges Bank.

There is a general consensus that connectivity between individual marine reserves should be factored in and, where possible, reserves should be designed as components of coherent and interconnected networks. Such networks are likely to offer improved performance, particularly in terms of resilience to changing environmental conditions. While short distance dispersers are likely to persist in almost all MPAs, long distance dispersers may be highly sensitive to reserve spacing and network design. Connectivity issues such as source-sink dynamics and larval dispersal distances (Cowen *et al.*, 2006) are therefore important considerations.

### 2. Benefits within MPA boundaries, depending on size and time

Fully protected marine reserves, where all take is prohibited, can have rapid and lasting benefits in terms of density, biomass, average organism size, and diversity inside reserves boundaries (Halpern and Warner, 2002). Halpern (2003) reviewed the performance of 89 MPAs at least partly closed to fishing, assessing four biological measures of recovery for each: density, biomass, size and diversity of species. All four were significantly greater inside than outside reserves for carnivorous fishes, herbivorous fishes and planktivorous fishes; with 90 per cent of the MPAs exhibiting increased biomass, 63 per cent increased numbers, and 59 per cent increased average size of species. Invertebrate density and diversity (but not invertebrate biomass and size) were also found to be greater inside than outside reserves.

Roberts (2007) suggests that these are minimum expectations for short term performance given that they aggregate results across species and degrees of MPA protection, and that results for no-take areas and well enforced MPAs will be even more significant. However a synthesis of coastal no-take reserve studies in 31 temperate and tropical locations found that only previously fished species showed overall enhanced abundances, with increasing positive effects on top trophic levels across time, and predation and competition suppressing 19 per cent of species on average (Micheli *et al.*, 2004).

Recovery of depleted stocks can be seen in as little as 2–5 years following establishment of no-take areas (Gell and Roberts, 2003). Refer to Case Study 2: St Lucia, for an example. However longer term protection may be required for population recoveries in many cases (Russ and Alcala, 2004). A review of 19 European marine reserves shows greater benefits for commercial fish species and species richness the longer protection has been in place (Claudet *et al.*, 2008).

## Case Study 2: St Lucia Soufrière Management Area

The Soufrière Marine Management Area (SMMA) was created in 1995 along the southwest coast of the Caribbean island of St Lucia. A network of five marine reserves along 11 km of coast, it covers about 35 per cent of the coral reef fishing grounds, and was designed to rehabilitate the severely overexploited reef fishery.

The marine reserves had a rapid impact on reef fish populations. Five exploited fish families tripled their biomass within five years, with a doubling in adjacent fishing grounds. This was despite the 35 per cent decrease in fishing grounds and redirection of fishing effort away from reserves. In the following two years, biomass remained consistent, with further increases likely obscured by damage to reefs from Hurricane Lenny in late 1999.

These results have been critiqued as too rapid to be due to the reserve and specific to coral reef fisheries. The speed of rebound in this case may reflect the fact that the St Lucia fishery mainly targets small, short-lived, rapidly growing species, and the close proximity of fishing grounds to reserves).

Despite the contention in the peer-reviewed literature, interviews with local fishers showed that most felt better off with reserves than without. Younger fishers were especially positive about the benefits. Interviews and boat surveys also revealed that growing numbers of fishers were visiting the SMMA from villages to the north and south. This suggests that the benefits were a result of the Soufrière reserves rather than being due to a regional oceanographic shift that increased recruitment.

Sources: Hilborn (2002); Roberts *et al.* (2001, 2002); Tupper *et al.* (2002).

A greater proportion of individuals may be expected to live to reproduce inside a no-take zone, and individuals grow to larger sizes over time. Larger fish in many cases produce more and 'better' offspring than smaller ones, enhancing the effects of reserves on population recovery. A 22-pound grouper for example produces 4.5 times the eggs (per pound) as a one pounder (Roberts, 2007). In addition, reserves can foster natural, extended age structures that help populations persist (Roberts *et al.*, 2005). Because more eggs and larvae are produced inside a no-take area with increased years of protection, recovery can also accelerate over time. Gell and Roberts (2003) report examples of the density or catch per unit effort of commercially important species being five or more times higher after just over a decade of well enforced protection in no-take areas.

While any no-take protection may increase fish density and diversity (Claudet *et al.*, 2008), the size of reserves also appears to be critical to meet fisheries recovery goals. If reserves are small and most individuals range in and out of no take areas, capture rates may exceed any enhanced reproductive output. However, in large reserves, a greater fraction of wider ranging species remain protected and species self-recruit (Claudet *et al.*, 2008).

## 3. Benefits outside MPA boundaries

A central goal of MPAs from a fisheries management perspective is 'spillover' across the no-take boundary, enriching the surrounding fishery, and improving its sustainability. Such spillover may involve the net movement of individuals, or eggs and larvae, from inside to outside the reserve boundaries. Net movement of animals may be a density dependent process, as individuals move outside reserve boundaries in search of new habitat and resources to avoid competition or predation (Almany *et al.*, 2007; Jones *et al.*, 2007). Movement of eggs and larvae is largely passive, driven by the physical movement of water masses (currents and tides) (Almany *et al.*, 2009).

Spillover effects therefore depend strongly on the characteristics of both the species in question (including life history and movement patterns), the physical properties of the ecosystem (currents and tides) (Gell and Roberts, 2003), and the design of the MPA itself (Forcada *et al.*, 2008). The proportion of reserve to exploited area, edge versus area ratio, transfer rates of fish across the boundary, fishing rates outside over time, edge permeability and the age of the reserve (references outlined in McClanahan and Mangi, 2000) all affect spillover. Spillover benefits have proved difficult to demonstrate empirically (Lundquist and Granek, 2005), but are well reported anecdotally. Refer to Case Study 3 for a Kenyan study on factors that may limit spillover.

Buffer zones around no-take areas are attractive to artisanal fishers, potentially increasing the fishing pressure in these areas (more than if they were unprotected with fishers mainly 'fishing the line' of the reserve edge) (Stelzenmuller *et al.*, 2007). This may explain the fact that larger buffer zones are associated with decreased commercial fish recovery within no-take areas (Claudet *et al.*, 2008).

Perhaps the most serious challenge for the design of MPA networks that have the best chance of favourably contributing to fisheries sustainability and yield is the paucity of empirical information on how they work at the population level. Reviews of why attributes such as biomass, mean age and diversity increase for some species but not others (e.g. Micheli *et al.*, 2004) are rarer than studies that consolidate many species (e.g. Halpern, 2003). Furthermore, effects on fishery yields have been reviewed for only a small fraction of MPAs.

MPAs can generate areas with high lifetime egg production, increasing stock recruitment and sustainability but not necessarily enhanced fisheries yield. Hart (2006) points out that comparisons between yields with and without marine reserves depend on the definition of 'yield'. Number-based yield calculations often used in marine reserve models (e.g.

Hastings and Botsford 1999) tend to favour marine reserves more than those which calculate yield in terms of weight.

### Case study 3: Mombasa Marine National Park, Kenya

The Mombasa Marine Park was established in 1987, with enforcement from late 1991. It covers 6.2 square kilometres of nearshore coast, 50–60 per cent of the original fishing ground. All fishing is excluded. Resources in such areas are typically heavily exploited by artisanal fishers, with reproductively active fish gone.

Follow up studies signal that 6–7 years on, spillover from the park is barely enhancing fisheries catch in adjacent areas. Fish weights increased inside the no-take area, however total fish catches in adjacent areas still did not match pre-park levels. Catch levels did however stabilise, and trap catches (per unit area and per fisher) immediately adjacent to the park increased. McClanahan and Mangi (2000) hypothesise that the ratio of protected area to fishing ground may be too high, reproductively active fish offshore may have a greater impact on larval recruitment to the fishery than those in the park, or it may be too early to see significant spillover.

Sources: McClanahan and Kaunda-Arara (1996); McClanahan and Mangi (2000).

Even if reserves do not increase fisheries yield, there is evidence that, if large enough, reserves improve overall sustainability. Simulation models can help elucidate how the fraction of a fish habitat assigned to MPAs affects the sustainability of take in surrounding fisheries (Mangel, 1998). Doyen and Bene (2003) constructed a model that suggested reserves covering at least 25 per cent of fisheries ensure the viability of the stocks above the safe minimum biomass level, even if they do not increase yield. They cite Mangel (2000), who in contrast to Hastings and Botsford (1999), concludes that catch declines with increased reserve size, but reserves decrease catch variability mitigating against 'boom and bust'. Doyen and Bene (2003) also suggest that optimal reserve size can be calculated for a guaranteed catch level. Crucial population parameters not incorporated into their model include the target species' use of space and life stage at harvest.

## Resilient Ecosystems and Fisheries

### 1. Resilience

Ecosystem resilience (coined by Holling, 1973) is "the extent to which ecosystems can absorb recurrent natural and human perturbations and continue to regenerate without slowly degrading or unexpectedly flipping into alternate states" (Hughes *et al.*, 2005). Disturbance of foodweb structure, primarily through fishing's removal of large apex predators and herbivores, has undermined the resilience of many marine systems (Worm *et al.*, 2006), as has habitat degradation (Halpern *et al.*, 2007). Other human-induced stressors such as pollution, sedimentation and temperature increases from climate change can also damage habitat, or exacerbate degradation from fishing.

## 2. Managing for resilience in fisheries

Managing for resilience, and other measures of community stability, has become a top priority for contemporary planning, management and conservation strategies (Hughes *et al.*, 2007; Pressey *et al.*, 2007). We suggest that managing for resilience should also be a goal for fisheries management, as resilient marine ecosystems will have the greatest capacity to continue to yield sustainable catches in the face of environmental change.

Careful marine spatial planning (following principles of 'integrated coastal zone management') is a crucial part of the solution. Even though all external stressors cannot be removed by spatial planning, reducing those impacts that can be managed reduces the overall synergistic impact of stressors, maximising the chance that marine systems can remain productive and function in the face of changing conditions (Pressey *et al.*, 2007). Maintaining resilient ecosystems with healthy populations of all constituent species also provides the potential to shift target species if unforeseen events reduce the populations or catchability of species previously targeted. It is a direct application of the precautionary principle. In terrestrial conservation planning this has been referred to as the 'portfolio effect' (Tilman, 1999), where a more diverse array of species fulfil overlapping ecological functions.

## 3. MPA contribution to resilience

MPAs, and especially no-take reserves, contribute to resilience by providing areas where communities have the best possible chance of: (a) recovering from over-exploitation; (b) adapting to changing environmental conditions; and (c) resisting diffuse and/or synergistic impacts such as climate change and pollution (Day, 2006). No-take reserves and MPAs under ecosystem-based management protect all species in the area equally whether they have been judged important or not to date; 'unknown' species and all (Nowlis and Friedlander, 2005). 'Unknowns' may have critical roles in supporting system resilience.

## 4. Adaptive management

Management strategies that aim for resilience should ideally be flexible and adaptive, using results from monitoring to refine management. Adaptive management is a way to deal with uncertainty in natural resource decision making. The core concept is that policy choices should be treated as deliberate, large-scale experiments (Walters, 2007). Adaptive management explicitly recognises the complex dynamics, thresholds and uncertainties of marine ecosystems, and multiple possible management outcomes (Hughes *et al.*, 2005). Such flexibility allows management strategies to navigate changes in governance systems and be updated with improved understandings of resource and ecosystem dynamics (Hughes *et al.*, 2005; Pressey *et al.*, 2007).

This adaptive approach to management contrasts with a 'hands off' school of thought, which trusts in the regenerative capacity of ecosystems - that they will move towards more complex resilient systems over the long term without ongoing management 'tweaks'. However, experience to date suggests that, while 'rules of thumb' should be used to guide expedient establishment of reserves in the first instance, continual refinement of management approaches with improving knowledge will yield management strategies with the highest performance (Walters, 2007).

Nevertheless, more MPAs fail to achieve objectives because of inadequate enforcement than a poor design-ecosystem match (Roberts, 2007). Adaptive management supported by scientific quality monitoring is also difficult in developing countries with insufficient infrastructure and resources.

Furthermore, with the current low percentage area of ocean protection, and high rate of deterioration, even if a good fraction of MPAs achieve their protection or restoration objectives, fisheries are likely to continue to decline on local, regional and global scales. This suggests that in the immediate term, rather than case-by-case adaptive management, our principal focus should be establishing enough MPAs to restore sustainable fisheries at national, regional and global scales, building in generic principles of good ecological MPA design on set-up and ongoing monitoring wherever possible.

## Conclusions

Because of severe fisheries depletion, the degradation of marine ecosystem services worldwide, and a growing awareness of the importance of marine ecosystem resilience, the fisheries management tide is turning towards protection and restoration. Ecosystem-based management is the favoured approach, and represents a shift in focus from individual species to ecosystems. Space-based approaches are the most tenable way to achieve EBM, and MPAs (including no-take areas) represent the core of such a system. However, MPAs are unlikely to meet fisheries sustainability or ecosystem conservation goals on their own. Sustainable fisheries management, as a component of EBM, requires the urgent and complementary implementation of both MPAs, and effort, gear and take restrictions in fished areas. In other words layered management regimes.

Well-designed MPAs can increase fish yields in surrounding areas. The density, biomass, size and diversity of species (in particular those previously fished) is enhanced inside MPAs, and surrounding fisheries may benefit through spillover of individuals, eggs or larvae. Active enforcement of MPAs is generally critical to such catch enhancement. In addition to increasing fisheries yields, MPAs can also increase sustainability by ensuring stocks remain viable, guaranteeing some ongoing catch flows. However, the fisheries enhancement (and ecosystem protection) efficacy of MPAs is strongly dependent on their design, placement, and spatial and temporal scale.

MPAs need to be carefully sized and located to meaningfully enhance recovery, sustainability and resilience. Large reserves are generally more effective, and carefully planned networks designed with connectivity in mind stand the best chance of success.

The protection of key habitats such as spawning and nursery grounds from extraction is very valuable, especially overlapping areas used by multiple species. As a rule of thumb, destructive fishing methods that degrade habitat should be prevented.

The benefit of reserves to surrounding fisheries depends on the proportion of protection versus exploited area, edge versus area ratio, transfer rates of fish across the boundary, fishing rates outside over time, edge permeability and the length of time protection is in place. MPA effects generally seem to accelerate over time. Initial recovery may be slow because of the harm already done to populations and ecosystems, and some fisheries may be irreversibly damaged. Long-lived species can be very slow to recover. Nevertheless, recovery may be observable in coral reef fisheries in as little as 2–5 years.

By enhancing sustainability of supply or yields into the future, MPAs offer economic and social benefits to fishers, yet their establishment can be socially controversial. Collaborative involvement of users in the design process, as well as full communication of the anticipated benefits is critical. Participatory process fosters ownership and self enforcement.

Supporting ecosystems should be a key goal for fisheries management in Commonwealth countries and worldwide. Greater ecosystem resilience enhances the prospect of ongoing, sustainable catch in the face of environmental change. The widespread establishment of MPAs is an essential strategy for the recovery of marine stocks, habitats and ecosystems. Commonwealth countries can lead the way with MPAs as a central component of layered management regimes, ensuring fishery and ecosystem sustainability into the future.

## References

Agardy, T., P. Bridgewater, M. P. Crosby, J. Day, P. K. Dayton, R. Kenchington, D. Laffoley, P. McConney, P. A. Murray, J.E. Parks, and L. Peau (2003) Dangerous targets? Unresolved issues and ideological clashes around marine protected areas. *Aquatic Conservation-Marine and Freshwater Ecosystems* 13, pp. 353–367.

Almany, G.R., M.L. Berumen, S.R. Thorrold, S. Planes, and G.P. Jones (2007) Local replenishment of coral reef fish populations in a marine reserve. *Science* 316, pp. 742–744.

Almany, G.R., S.R. Connolly, D.D. Heath, J.D. Hogan, G.P. Jones, L.J. McCook, M. Mills, R.L. Pressey, and D.H. Williamson (2009) Connectivity, biodiversity conservation and the design of marine reserve networks for coral reefs. *Coral Reefs* 28 (2), pp. 339–351.

Balmford, A., P. Gravestock, N. Hockley, C.J. McClean, and C.M. Roberts (2004) The worldwide costs of marine protected areas. *Proceedings of the National Academy of Sciences of the United States of America* 101, pp. 9694–9697.

Bartlett, C.Y., K. Pakoa, and C. Manua (2009) Marine reserve phenomenon in the Pacific islands. *Marine Policy* 33, pp. 673–678.

Baskett, M.L., S.A. Levin, S.D. Gaines, and J. Dushoff (2005) Marine reserve design and the evolution of size at maturation in harvested fish. *Ecological Applications* 15, pp. 882–901.

Berkes, F., Mahon, R., McConney, P., Pollnac, R., Pomeroy, R. (2001) Managing small-scale fisheries: alternative directions and methods (Ottawa: International Development Research Centre).

Botsford, L.W., F. Micheli, and A. Hastings (2003) Principles for the design of marine reserves. *Ecological Applications* 13, pp. S25–S31.

Carr, M.H., J.E. Neigel, J.A. Estes, S. Andelman, R.R. Warner, and J.L. Largier (2003) Comparing marine and terrestrial ecosystems: Implications for the design of coastal marine reserves. *Ecological Applications* 13, pp. S90–S107.

Cinner, J.E., and S. Aswani (2007) Integrating customary management into marine conservation. *Biological Conservation* 140, pp. 201–216.

Claudet, J., C.W. Osenberg, L. Benedetti-Cecchi, P. Domenici, J.A. Garcia-Charton, A. Perez-Ruzafa, F. Badalamenti, J. Bayle-Sempere, A. Brito, F. Bulleri, J.M. Culioli, M. Dimech, J.M. Falcon, I. Guala, M. Milazzo, J. Sanchez-Meca, P. J. Somerfield, B. Stobart, F. Vandeperre, C. Valle, and S. Planes (2008) Marine reserves: size and age do matter. *Ecology Letters* 11, pp. 481–489.

Costanza, R., R. dArge, R. deGroot, S. Farber, M. Grasso, B. Hannon, K. Limburg, S. Naeem, R.V. Oneill, J. Paruelo, R.G. Raskin, P. Sutton, and M. vandenBelt (1997) The value of the world's ecosystem services and natural capital. *Nature* 387, pp. 253–260.

Cowen, R.K., C.B. Paris and A. Srinivasan (2006) Scaling of connectivity in marine populations. *Science* 311, pp. 522–527.

Crowder, L.B., E.L. Hazen, N. Avissar, R. Bjorkland, C. Latanich, and M.B. Ogburn (2008) The impacts of fisheries on marine ecosystems and the transition to ecosystem-based management. *Annual Review of Ecology Evolution and Systematics* 39, pp. 259–278.

Day, J. (2006) Marine protected areas. In M. Lockwood and A. Kothari (eds) *Managing protected areas: a global guide*. (London: Earthscan), pp. 606–634.

Dayton, P. K., E. Sala, M. J. Tegner, and S. Thrush (2000) Marine reserves: Parks, baselines, and fishery enhancement. *Bulletin of Marine Science* 66, pp. 617–634.

Doyen, L., and C. Bene C. (2003) Sustainability of fisheries through marine reserves: a robust modeling analysis. *Journal of Environmental Management* 69, pp. 1–13.

Edren, S.M.C., and S.H. Gruber. (2005) Homing ability of young lemon sharks, Negaprion brevirostris. *Environmental Biology of Fishes* 72, pp. 267–281.

FAO. (2009) *The State of the World Fisheries and Aquaculture 2008*. Food and Agriculture (Rome: FAO).

Fernandes, L., J. Day, A. Lewis, S. Slegers, B. Kerrigan, D. Breen, D. Cameron, B. Jago, J. Hall, D. Lowe, J. Innes, J. Tanzer, V. Chadwick, L. Thompson, K. Gorman, M. Simmons, B. Barnett, K. Sampson, G. De'ath, B. Mapstone, H. Marsh, H. Possingham, I. Ball, T. Ward, K. Dobbs, J. Aumend, D. Slater and K. Stapleton. (2005) Establishing representative no-take areas in the Great Barrier Reef: Large-scale implementation of theory on marine protected areas. *Conservation Biology* 19, pp. 1733–1744.

Forcada, A., J.T. Bayle-Sempere, C. Valle, and P. Sanchez-Jerez (2008) Habitat continuity effects on gradients of fish biomass across marine protected area boundaries. *Marine Environmental Research* 66, pp. 536–547.

GBRMPA. (2003) *Summary Report of the Social and Economic Impacts of the Re-zoning of the Great Barrier Reef Marine Park*. Great Barrier Reef Marine Park Authority, Townsville. *http://www.gbrmpa.gov.au*

Gell, F.R., and C.M. Roberts (2003) Benefits beyond boundaries: the fishery effects of marine reserves. *Trends in Ecology & Evolution*, pp. 18: 448–455

Halpern, B. S. (2003) The impact of marine reserves: Do reserves work and does reserve size matter? *Ecological Applications* 13 pp. S117–S137.

Halpern, B.S., K.L. McLeod, A.A. Rosenberg, and L.B. Crowder. (2008) Managing for cumulative impacts in ecosystem-based management through ocean zoning. *Ocean & Coastal Management* 51, pp. 203–211.

Halpern, B.S., H.M. Regan, H.P. Possingham, and M.A. McCarthy. (2006) Accounting for uncertainty in marine reserve design. *Ecology Letters* 9, pp. 2–11.

Halpern, B.S., K.A. Selkoe, F. Micheli, and C.V. Kappel (2007) Evaluating and ranking the vulnerability of global marine ecosystems to anthropogenic threats. *Conservation Biology* 21, pp. 1301–1315.

Halpern, B.S., and R.R. Warner (2002) Marine reserves have rapid and lasting effects. *Ecology Letters* 5, pp. 361–366.

Hart, D.R. (2006) When do marine reserves increase fisheries yield? *Canadian Journal of Fisheries and Aquatic Sciences* 63, pp. 1445–1449.

Hastings, A. and L.W. Botsford (1999) Equivalence in yield from marine reserves and traditional fisheries management. *Science* 284, pp. 1537–1538.

Hilborn, R. (2002) Marine reserves and fisheries management. *Science* 295, pp. 1233–1235.

Holling, C.S. (1973) Resilience and Stability of Ecological Systems. *Annual Review of Ecology and Systematics* 4, pp. 1–23.

Hooker, S.K. and L.R. Gerber. (2004) Marine reserves as a tool for ecosystem-based management: the potential importance of megafauna. *BioScience* 54, pp. 27–39.

Hughes, T.P., D.R. Bellwood, C. Folke, R.S. Steneck, and J. Wilson. (2005) New paradigms for supporting the resilience of marine ecosystems. *Trends in Ecology & Evolution* 20, pp. 380–386.

Hughes, T. P., L. H. Gunderson, C. Folke, A. H. Baird, D. Bellwood, F. Berkes, B. Crona, A. Helfgott, H. Leslie, J. Norberg, M. Nystrom, P. Olsson, H. Osterblom, M. Scheffer, H. Schuttenberg, R.S. Steneck, M. Tengoe, M. Troll, B. Walker, J. Wilson, and B. Worm. (2007) Adaptive management of the great barrier reef and the Grand Canyon world heritage areas. *Ambio* 36, pp. 586–592.

Hutchings, J. A. (2000) Collapse and recovery of marine fishes. *Nature* 406, pp. 882–885.

Jackson, J.B.C., M.X. Kirby, W.H. Berger, K.A. Bjorndal, L.W. Botsford, B.J. Bourque, R.H. Bradbury, R. Cooke, J. Erlandson, J.A. Estes, T. P. Hughes, S. Kidwell, C.B. Lange, H.S. Lenihan, J. M. Pandolfi, C.H. Peterson, R.S. Steneck, M.J. Tegner, and R.R. Warner. (2001) Historical overfishing and the recent collapse of coastal ecosystems. *Science* 293, pp. 629–638.

Jones, G. P., M. Srinivasan, and G. R. Almany (2007) Population Connectivity and Conservation of Marine Biodiversity. *Oceanography* 20, pp. 100–111.

Jones, P.J.S. (2001) Marine protected area strategies: issues, divergences and the search for middle ground. *Reviews in Fish Biology and Fisheries* 11, pp. 197–216.

Law, R. (2007) Fisheries-induced evolution: present status and future directions. *Marine Ecology-Progress Series* 335, pp. 271–277.

Lotze, H.K., and B. Worm (2009) Historical baselines for large marine animals. *Trends in Ecology & Evolution* 24, pp. 254–262.

Lundquist, C.J., and E.F. Granek. (2005) Strategies for successful marine conservation: Integrating socioeconomic, political, and scientific factors. *Conservation Biology* 19, pp. 1771–1778.

Mangel, M. (1998) No-take areas for sustainability of harvested species and a conservation invariant for marine reserves. *Ecology Letters* 1, pp. 87–90.

Margules, C.R., and R.L. Pressey (2000) Systematic conservation planning. *Nature* 405, pp. 243–253.

McClanahan, T.R. and S. Mangi (2000) Spillover of exploitable fishes from a marine park and its effect on the adjacent fishery. *Ecological Applications* 10, pp. 1792–1805.

McClanahan, T.R. and B. Kaunda-Arara. (1996) Fishery recovery in a coral-reef marine park and its effect on the adjacent fishery. *Conservation Biology* 10, pp.1187–1199.

Micheli. F., Halpern, B.S., Botsford, L.W., Warner, R.R. (2004) Trajectories and correlates of community change in no-take marine reserves. *Ecological Applications* 14, pp. 1709–1723.

Myers, R. A., and B. Worm (2003) Rapid worldwide depletion of predatory fish communities. *Nature* 423, pp. 280–283.

Norse, E., and L.B. Crowder (2005) *Marine Conservation Biology: The Science of Maintaining the Sea's Biodiversity* (Washington: Island Press).

Nowlis, J.S., and A.M. Friedlander (2005) Marine reserve design and function for fisheries management in E. Norse, and L. Crowder, editors. *Marine Conservation Biology: the Science of Maintaining the Sea's Biodiversity* (Washington: Island Press).

Palumbi, S.R., S.D. Gaines, H. Leslie, and R.R. Warner (2003) New wave: high-tech tools to help marine reserve research. *Frontiers in Ecology and the Environment* 1, pp. 73–79.

Pauly, D., V. Christensen, V. Dalsgaard, R. Froese, and J. Torres, R. (1998) Fishing down marine food webs. *Science* 279, pp. 860–863.

Pinnegar, J.K., and G. H. Engelhard (2008) The 'shifting baseline' phenomenon: a global perspective. *Reviews in Fish Biology and Fisheries* 18, pp. 1–16.

Pressey, R.L., M. Cabeza, M.E. J. Watts, R.M. Cowling, and K.A. Wilson (2007) Conservation planning in a changing world. *Trends in Ecology & Evolution* 21, pp. 583–592.

Roberts, C. (2007) *The Un-Natural History of the Sea.* (London: Island Press).

Roberts, C.M., S. Andelman, G. Branch, R.H. Bustamante, J.C. Castilla, J. Dugan, B.S. Halpern, K.D. Lafferty, H. Leslie, J. Lubchenco, D. McArdle, H.P. Possingham, M. Ruckelshaus, and R.R. Warner (2003). Ecological criteria for evaluating candidate sites for marine reserves. *Ecological Applications* 13, pp. S199–S214.

Roberts, C.M., J.A. Bohnsack, F. Gell, J. F. Hawkins, and R. Goodridge. (2001) Effects of marine reserves on adjacent fisheries. *Science* 294, pp. 1920–1923.

Roberts, C.M., J.A. Bohnsack, F. Gell, J. P. Hawkins, and R. Goodridge. (2002) Marine reserves and fisheries management – Response. *Science* 295, pp. 1234–1235.

Roberts, C.M., and J.P. Hawkins (1999) Extinction risk in the sea. *Trends in Ecology & Evolution* 14, pp. 241–246.

Roberts, C.M., J.P. Hawkins, and F. R. Gell (2005) The role of marine reserves in achieving sustainable fisheries. *Philosophical Transactions of the Royal Society B-Biological Sciences* 360, pp. 123–132.

Robichaud, D., and G.A. Rose (2004) Migratory behaviour and range in Atlantic cod: inference from a century of tagging. *Fish and Fisheries* 5, pp. 185–214.

Rodrigues, A.S.L., S.J. Andelman, M.I. Bakarr, L. Boitani, T.M. Brooks, R.M. Cowling, L.D.C. Fishpool, G. A. B. da Fonseca, K. J. Gaston, M. Hoffmann, J. S. Long, P. A. Marquet, J. D. Pilgrim, R.L. Pressey, J. Schipper, W. Sechrest, S.N. Stuart, L.G. Underhill, R.W. Waller, M.E.J. Watts, and X. Yan. (2004) Effectiveness of the global protected area network in representing species diversity. *Nature* 428, pp. 640–643.

Ruckelshaus, M., T. Klinger, N. Knowlton, and D.P. DeMaster (2008) Marine ecosystem-based management in practice: Scientific and governance challenges. *BioScience* 58, pp. 53–63.

Russ, G.R., and A.C. Alcala (1998) Natural fishing experiments in marine reserves 1983–1993: community and trophic responses. *Coral Reefs* 17, pp. 383–397.

Russ, G.R., and A.C. Alcala (2004) Marine reserves: long-term protection is required for full recovery of predatory fish populations. *Oecologia* 138, pp. 622–627.

Sala, E., and N. Knowlton (2006) Global marine biodiversity trends. *Annual Review of Environment and Resources* 31, pp. 93–122.

Sarkar, S., R.L. Pressey, D.P. Faith, C.R. Margules, T. Fuller, D.M. Stoms, A. Moffett, K.A. Wilson,

K. J. Williams, P.H. Williams, and S. Andelman (2006) Biodiversity conservation planning tools: Present status and challenges for the future. *Annual Review of Environment and Resources* 31, pp. 123–159.

Smith, S.E., D.W. Au, and C. Show (1998) Intrinsic rebound potentials of 26 species of Pacific sharks. *Marine and Freshwater Research* 49, pp. 663–678.

Spalding, M.D., L. Fish, and L.J. Wood (2008) Toward representative protection of the world's coasts and oceans progress, gaps, and opportunities. *Conservation Letters*, 1, pp. 217–226.

Stefansson, G., and A.A. Rosenberg (2006) Designing marine protected areas for migrating fish stocks. *Journal of Fish Biology* 69, pp. 66–78.

Stelzenmuller, V., F. Maynou, and P. Martin (2007) Spatial assessment of benefits of a coastal Mediterranean Marine Protected Area. *Biological Conservation* 136, pp. 571–583.

Tilman, D. (1999) The ecological consequences of changes in biodiversity: a search for general principles. *Ecology* 80, pp. 1455–1474.

Trebilco, R., R. Gales, G.B. Baker, A. Terauds, and M.D. Sumner (2008) At sea movement of Macquarie Island giant petrels: Relationships with marine protected areas and Regional Fisheries Management Organisations. *Biological Conservation* 141, pp. 2942–2958.

Tupper, M.H., Wickstrom, Karl, Hilborn, Ray, C.M. Roberts, J.A. Bohnsack, F. Gell, J.P. Hawkins, and R. Goodridge (2002) Marine reserves and fisheries management. *Science* 295, pp. 1233–1235.

United Nations (2002) Report of the World Summit on Sustainable Development, New York. (*http://www.un.org/jsummit*).

Walters, C.J. (2007) Is adaptive management helping to solve fisheries problems? *Ambio* 36: pp. 304–307.

Watling, L., and E.A. Norse (1998) Disturbance of the seabed by mobile fishing gear: A comparison to forest clearcutting. *Conservation Biology* 12, pp. 1180–1197.

Wescott, G. (2006) The long and winding road: the development of a comprehensive, adequate and representative system of highly protected marine protected areas in Victoria, Australia. *Ocean and Coastal Management* 49, pp. 905–922.

White, C., and B.E. Kendall (2007) A reassessment of equivalence in yield from marine reserves and traditional fisheries managament. *Oikos* 116, pp. 2039–2043.

Wilson, J.A. (2006) Matching social and ecological systems in complex ocean fisheries. *Ecology and Society* 11:9. *http://www.ecologyandsociety.org/vol11/iss1/art9/*

Wood, L. J., and S. Dragicevic (2007) GIS-Based multicriteria evaluation and fuzzy sets to identify priority sites for marine protection. *Biodiversity and Conservation* 16, pp. 2539–2558.

Worm, B., E.B. Barbier, N. Beaumont, J. E. Duffy, C. Folke, B.S. Halpern, J. B.C. Jackson, H.K. Lotze, F. Micheli, S.R. Palumbi, E. Sala, K.A. Selkoe, J.J. Stachowicz, and R. Watson (2006) Impacts of biodiversity loss on ocean ecosystem services. *Science* 314, pp. 787–790.

Young, O.R., G. Osherenko, J. Ogden, L.B. Crowder, J.A. Wilson, J.C. Day, F. Douvere, C. N. Ehler, K.L. McLeod, B.S. Halperin, and R. Peach (2007) Solving the crisis in ocean governance: Place-based management of marine ecosystems. *Environment* 49, pp. 20–32.

# 14 Vulnerability and Sustainability of Fish Stocks Worldwide: With Emphasis on Fish Stocks within the Commonwealth

Rashid Sumaila and William W.L. Cheung

## Introduction

Fisheries resources in the world ocean are increasingly being over-exploited. After the Second World War, fishing effort increased rapidly. This was partly due to the mechanisation of fishing fleets, advances in fishing technology and methods, and ineffective fisheries management. Excessive fishing effort has led to large-scale depletion of marine populations and changes in ecosystem structure (e.g. Pauly *et al.*, 2002; Myers and Worm, 2003; Worm *et al.*, 2006; Berkes *et al.*, 2006). For example, fishing depleted large-bodied, predatory fishes that are particularly vulnerable. Fish catch becomes increasingly dominated by species lower in the food-chain and less vulnerable (more resilient) smaller fishes that could sustain high fishing pressure, a phenomenon known as 'fishing down marine foodweb' (Pauly *et al.*, 1998). Over-exploitation results in substantial socio-economic loss to the fishing industries and communities (World Bank and FAO, 2008). It also threatens the long-term survival of marine species, particularly those that are vulnerable to fishing (Hutchings and Reynolds, 2004). Some of the vulnerable fishes such as certain species of groupers, sharks and rays have been listed under the IUCN Red List of Endangered Species (*www.redlist.org*), with a few fish and inver-tebrates listed by CITES (the Convention on International Trade in Endangered Species). The long-term sustainability (ecological, economic and social) of exploitation of many fish stocks in the world is seriously in doubt (Pauly *et al.*, 2002; Worm *et al.*, 2006). Since the Exclusive Economic Zones (EEZs) of the 53 member states of the Commonwealth are distributed across all the major ocean basins of the world (Figure 1), it is not surprising that they face similar challenges to the rest of the world in the sustainability of their fish stocks.

Some species are intrinsically vulnerable to fishing (Cheung *et al.*, 2005; Reynolds *et al.*, 2005) as a result of their biology and ecology. These include species that are large-bodied, long-lived or late-maturing. Given similar fishing pressure, species with higher intrinsic vulnerabilities are more prone to over-exploitation and this has large implications for the sustainability of fish stocks and the effect of fishing on marine ecosystems and fish communities. Thus, it is important to evaluate the relative vulnerability of exploited species and their relationship with fishing practices.

**Figure 1: Map of Exclusive Economic Zones of Commonwealth member countries**

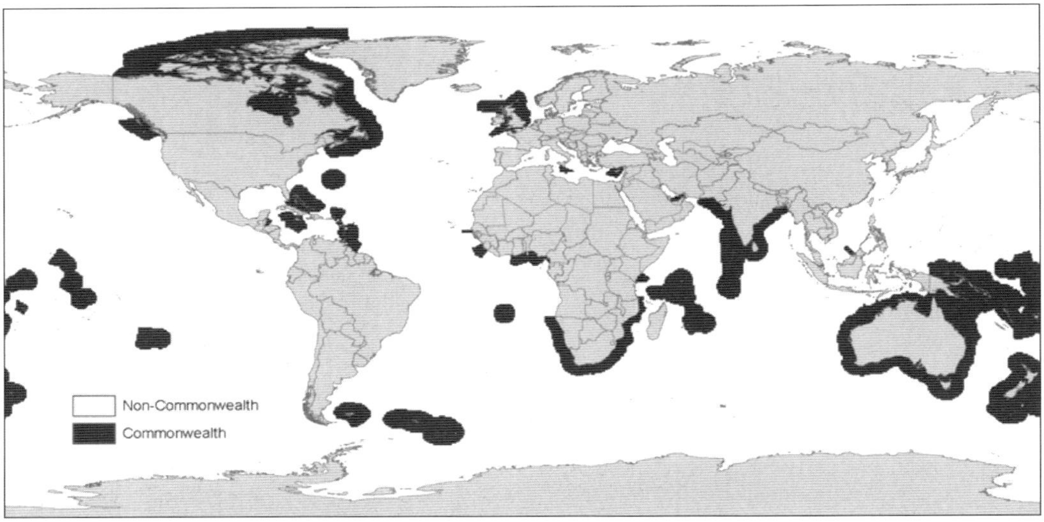


In this paper, we provide an overview of the vulnerability and sustainability of exploited fish stocks in the world ocean, with emphasis on the EEZs of member states of the Commonwealth. We use various published indicators to reveal the temporal and spatial trends and global data of reported catch as an indicator of the general status of sustainability of exploited fish stocks. Then, we evaluate the intrinsic vulnerability of fish stocks to fishing. Finally, we reveal the close relationship between the vulnerability of the organism to fishing and the sustainability of their exploitation over the past six decades.

## Sustainability of Exploited Fish Stocks

### 1. Data sources

We used the catch data from the *Sea Around Us* Project to evaluate the trends of reported global fish catch. The *Sea Around Us* Project developed an algorithm that disaggregated reported catch data from 1950 to 2004 into a 30' lat. x 30' lon. grid of the world ocean (see Watson *et al.*, 2004 and *www.seaaroundus.org* for details). The main source of catch data is the fisheries statistics from the Food and Agriculture Organisation of the United Nations (FAO), which is modified where appropriate with more reliable data. Thus, the *Sea Around Us* Project catch data allow us to investigate both the temporal trend and spatial patterns of change in global catch. Similarly, using the same dataset, we studied catches extracted from the EEZs of the member states of the Commonwealth over the past six decades.

### 2. Global catch trends

Global reported fish catch increased from less than 20 million tonnes in 1950 and peaked at around 80 million tonnes in the mid-1980s (Watson and Pauly, 2001) (Figure 2a). Up to 10 per cent of the global catch is contributed by the highly productive Peruvian anchovy (*Engraulis ringens*). When Watson and Pauly (2001) subtracted the Peruvian anchovy from the global catch, a steeper decline in global catch since the 1980s was demonstrated. Subdividing the catch into groups of species, catches of invertebrates and small pelagic fishes (<30 cm in total length) were relatively stable in the last two decades. However, catches of large pelagic (living in surface or midwater depth) and demersal (living on or near the seafloor) fishes declined consistently.

**Figure 2: Time-series of reported catch from 1950 to 2004: (a) globally and (b) from the Exclusive Economic Zones (EEZ) of members of the Commonwealth**

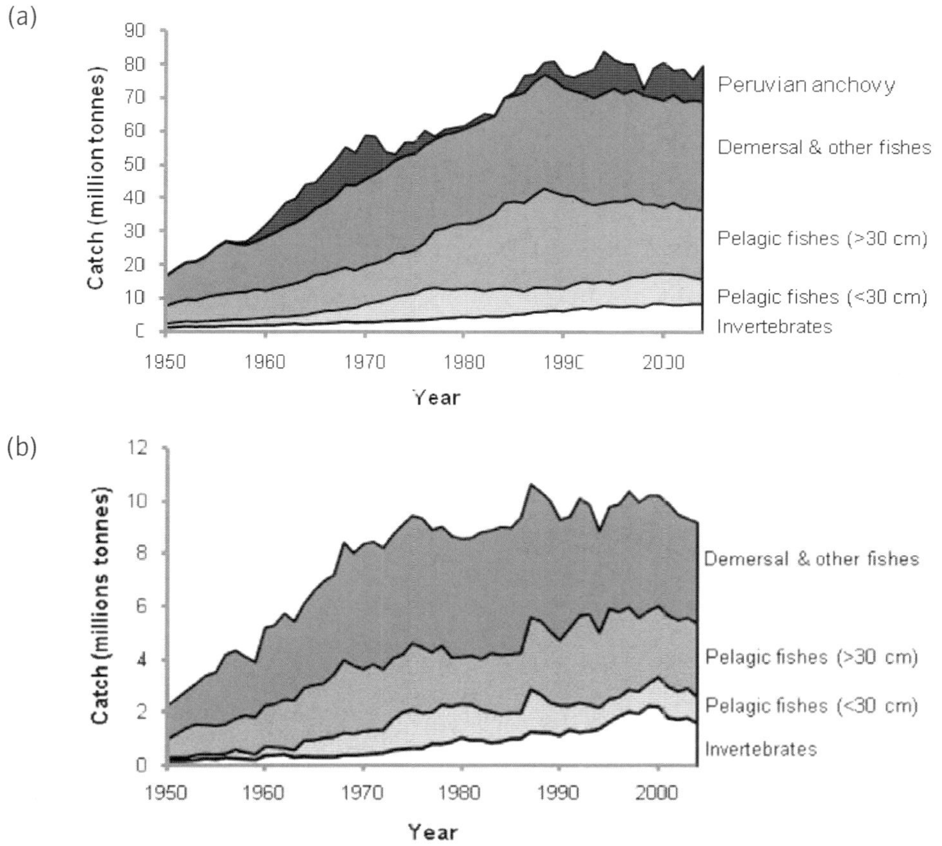


The global catch trend provides a general picture of the exploitation level of the world's fisheries resources. Since the 1980s, fishing effort in many regions continued to increase while there was little reduction in the overall fishing power. Thus, the reduction in catch (excluding the catch from Peruvian anchovy) indicates that the majority of the world's fisheries resources have already reached their maximum sustainable levels. Specifically, large predatory and demersal fishes show declining trends since the mid-1980s while there is no sign of a marked decrease in fishing effort and power of bottom trawlers; the fleet that mainly catches demersal species in the continental shelf. This suggests that these groups are already over-exploited.

Catch extracted from the EEZs of the member states of the Commonwealth reached a peak of around 9 million tonnes in the mid-1970s, a decade earlier than the global trend. The catch has then remained relatively stable (Figure 2b). Adjusting the statistics with potential mis-reported catch may result in an even steeper declining trend in global catch in recent decades (Watson and Pauly, 2001).

Evaluating reported catches by ocean basins showed that ocean basins with longer histories of industrial fishing reached their maximum catch potential as early as in the 1970s (Figure 3). For example, in the North Atlantic where large-scale industrial fishing started in the early 20th century, catch peaked at around 1970 and dropped consistently from then on. Similarly, the North Pacific, particularly the Bering Sea, was intensively fished by Russian, Japanese and US fleets. Catch from this region dropped continuously after reaching a peak in the 1980s.

**Figure 3: Catch time-series by ocean basins: North Atlantic, Central Atlantic, South Atlantic, Eastern Pacific, Western Pacific, Central East Pacific, Indo-Pacific, and South Pacific**

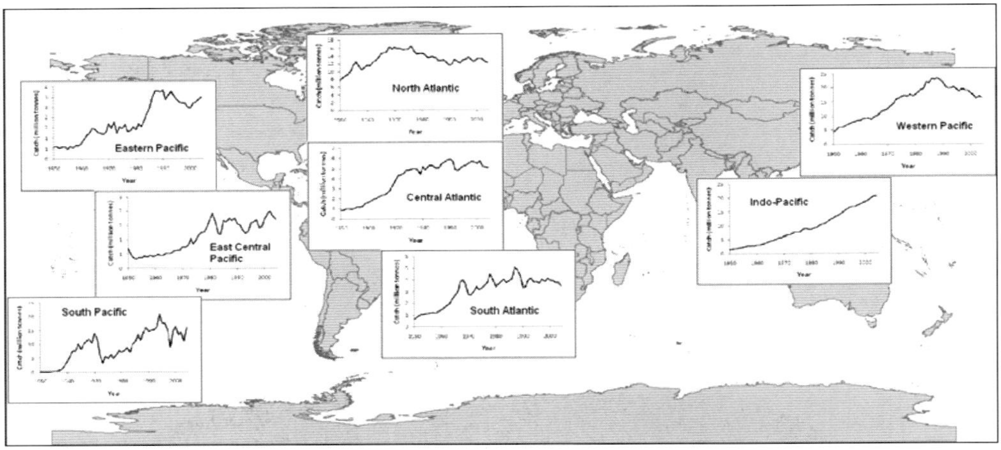


The continuous increase in catch from regions such as the Central East Pacific and Indo-Pacific regions does not imply that their fisheries are sustainable. Catch statistics from these regions are often poorly-reported. Particularly, the level of unreported or mis-reported catch is potentially high (Bailey *et al.*, 2008). Many fish stocks that are known to have been depleted by fishing (e.g. Humphead wrasse, *Cheilinus undulates*) are not reported in the national fisheries statistics or are reported as aggregated groups (e.g. other fish). Thus, the trends of reported catch in these regions may be providing only a conservative picture of the sustainability of the stocks. Based on the available evidence, correcting for the unreported or mis-reported catch may show a much more pessimistic trend in terms of the sustainability of the overall catch.

Spatially, we looked at the expansion of the peak catch (a proxy for fish stocks being fully-exploited) in the world ocean (Figure 4). In the 1950s, only a small proportion of the world ocean had reached their peak catch (fully-exploited). By the 1970s, catch peaked in most of the traditional fishing grounds for industrial fishing such as the North Atlantic and the Northeast Pacific. By the 2000s, fishing covered almost the whole world ocean. From these pictures, it is clear that the ocean has run out of refuge of unexploited fisheries resources.

**Figure 4: Area where maximum catch has been reached in each 30 x 30 minutes cell in (a) mid-1950s; (b) mid-1970s; and (c) mid-2000s**

(a) Mid-1950s

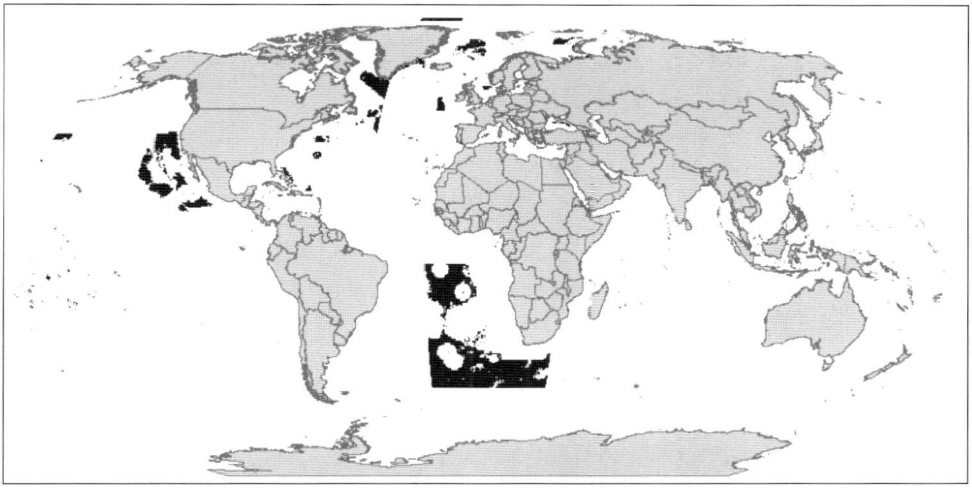

(b) Mid-1970s

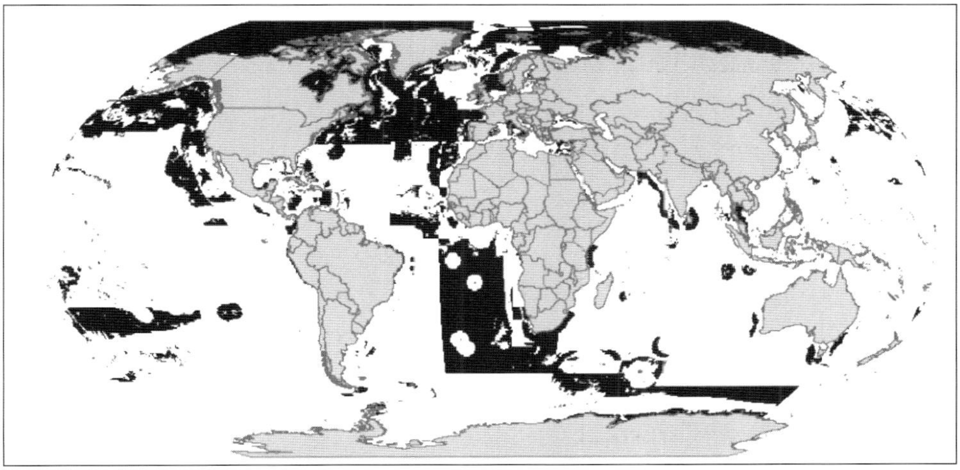

(c) Mid-2000s

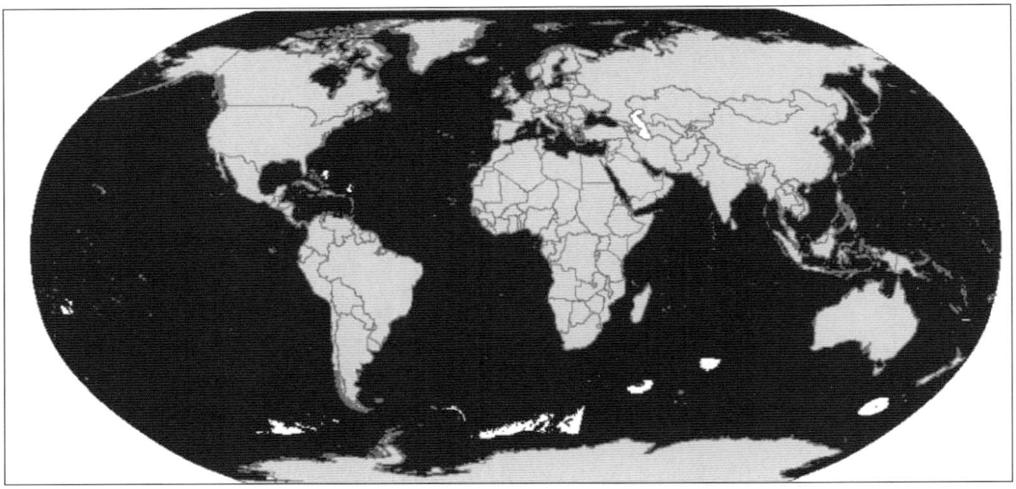

**Figure 5: Percentage of exploited stocks in the five status categories – under-developed, developing, fully-exploited, over-exploited and collapsed: (a) globally; and (b) within the EEZs of members of the Commonwealth**

(a) Global

(b) Within the EEZs of the members of the Commonwealth

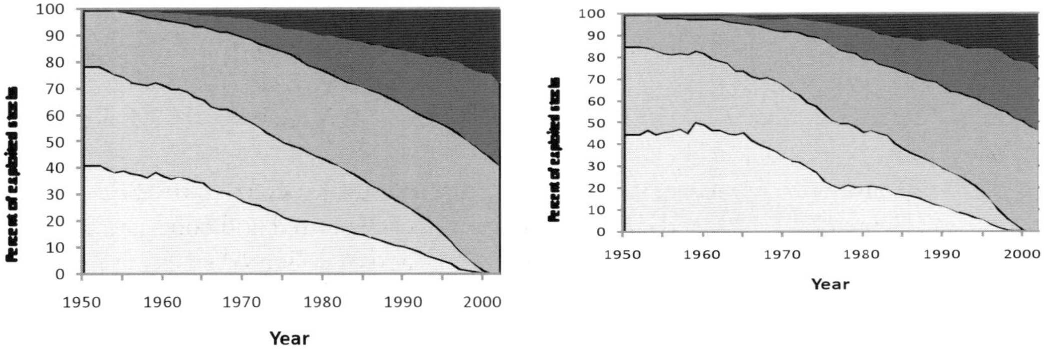


## 3. Status of exploited stocks

Based on catch time-series data and the categorisation schemes similar to those adopted by FAO (2006), we determined the exploitation status of an exploited fish stock. Here, we used each country's EEZ to delineate the boundary of a stock. Firstly, we determined the maximum catch of each time series. Secondly, following the approach adopted by FAO

(2006), we define various stages of exploitation of fish population. These stages include: 'under-developed' (low level of fishing effort, producing much lower catches of fish than the potential maximum level), 'developing' (rapid increase in fishing effort and catch), 'fully-developed' (reached maximum annual catch), 'over-exploited' (fish abundance is too low to maintain maximum level of catch) and collapsed (reduction of fish abundance by fishing to levels at which the catch is negligible compared to historical levels). For years before the maximum catch of a stock had been reached and the catch was below 10 per cent and 50 per cent of the maximum catch, a stock was considered 'under-developed' and 'developing', respectively. When annual catch exceeded 50 per cent of the maximum catch, a stock was considered 'fully-exploited'. For years after the maximum catch had been reached and annual catch was below 50 per cent of the maximum catch, the stock was considered 'over-exploited'. Finally, when catch declined to less than 10 per cent of the maximum catch, it was considered 'collapsed'.

Catch of some stocks fluctuate largely over time because of factors that are not primarily related to fishing (e.g. climate, market fluctuation). To avoid these extrinsic effects from confusing our analysis of the exploitation status of stocks, we applied a 5-year moving average to the catch time-series of each stock. Thus, the identified stock status would better reflect the long-term trend of the stocks. We evaluated the overall status of the exploited fish stocks globally, by country EEZs and, specifically for the member states of the Commonwealth, from 1950 to 2004.

Classifying exploited stocks (by country and taxon) into the above five exploitation ranks (under-developed, developing, fully-exploited, over-exploited, collapsed), the number of over-exploited and collapsed stocks in the world increased rapidly from less than 5 per cent in the late 1950s to over 50 per cent in the 2000s (Figure 5a). This means that catch from over half of the fish stocks decreased by over 50 per cent from their maximum level. There is virtually no exploited stock that is currently in the under-developed or developing stage. The pattern in the member states of the Commonwealth is similar to the global pattern.

The large proportion of over-exploited and collapsed stocks occurs in the EEZs of most countries (Figure 6). We plotted the proportion of over-exploited and collapsed stocks in each country. In the 1960s, the number of over-exploited or collapsed stocks in most part of the world is small, less than 10 per cent of their exploited fish stocks. However, by the 2000s, most EEZs in the northern hemisphere have high proportion of over-exploited or collapsed stocks. The small proportion of over-exploited or collapsed stocks in the Indo-Pacific region is possibly a result of the inaccurate catch statistics and has to be interpreted with caution.

**Figure 6: Percentage of stocks that are classified as over-exploited or collapsed in each EEZ: (a) in the 1960s and (b) in the 2000s**

(a) 1960s

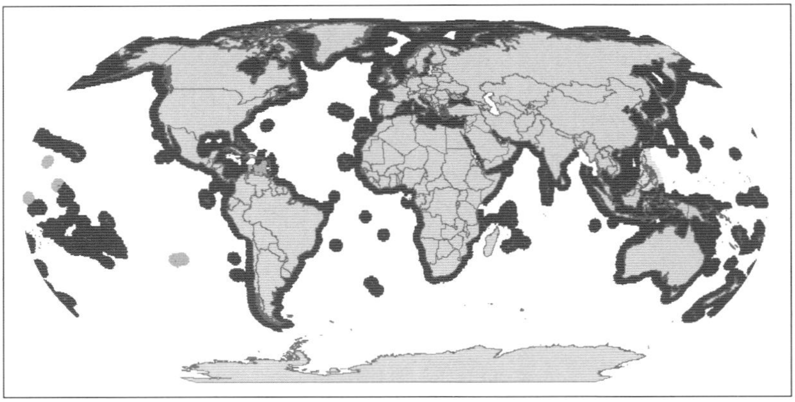

(b) 2000s

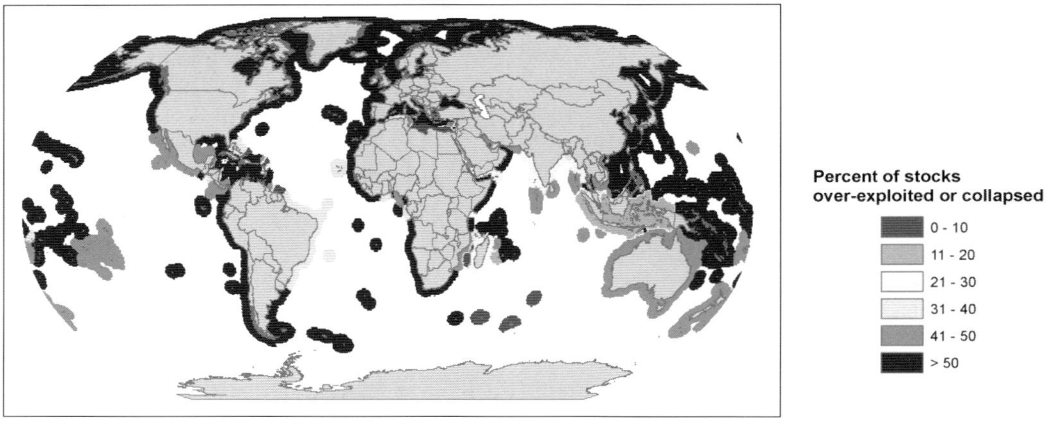


## Vulnerability of Fish Stocks

### 1. Intrinsic vulnerability index of fishes to fishing

An intrinsic vulnerability index was calculated, based on life history and ecological information extracted from FishBase (*www.fishbase.org*) (Cheung *et al.*, 2005). These data included maximum body length, age at maturity, longevity, growth rate, natural mortality rate, fecundity and fish's behavior in forming aggregations. The index is scaled from 1 to 100, with 100 being most vulnerable to fishing. For taxa that are reported as an aggregate group at genera or family levels rather than as a single species, we calculated their intrinsic vulnerability indices using the mean index value of species that belong to the same taxonomic groups.

We calculated the change in the average intrinsic vulnerability of the exploited fish stocks weighted by their annual catch over time (Cheung *et al.*, 2007). Given that species with higher intrinsic vulnerability would be depleted more rapidly than less vulnerable species, other factors being equal, one would expect that intensive fishing would result in change in fish community structure. In such a case, we should observe a decline in the proportion of the vulnerable species in the catch, and a resultant decrease in the mean intrinsic vulnerability of the catch over time. We use this as an indicator of fishing-induced changes in fish community structure.

## 2. Comparison of intrinsic vulnerability between fish communities

Comparing the intrinsic vulnerability of fish that are associated with different environments, we have shown that those fish that are closely associated to seamounts and other deep water habitats are particularly vulnerable to fishing (Figure 7). Based on information from FishBase (*www.fishbase.org*), the intrinsic vulnerability index of a sample of over 2,000 species of marine fish was calculated (Cheung *et al.*, 2007). The major habitat with which each species is mainly associated was identified, and the average intrinsic vulnerability of fish between these major habitat types was compared. Fishes that occur in seamounts, or aggregate around seamounts, are significantly more vulnerable than other fishes. In fact, their vulnerability is similar to those species that are listed under the IUCN Red List of Endangered Species. Intrinsic vulnerability indices of fish species also increase with the depth of their habitat (see Cheung *et al.*, 2007), so deep water fish tend to be more vulnerable to fishing.

**Figure 7: Mean intrinsic vulnerability index of marine fishes associated with different environments**

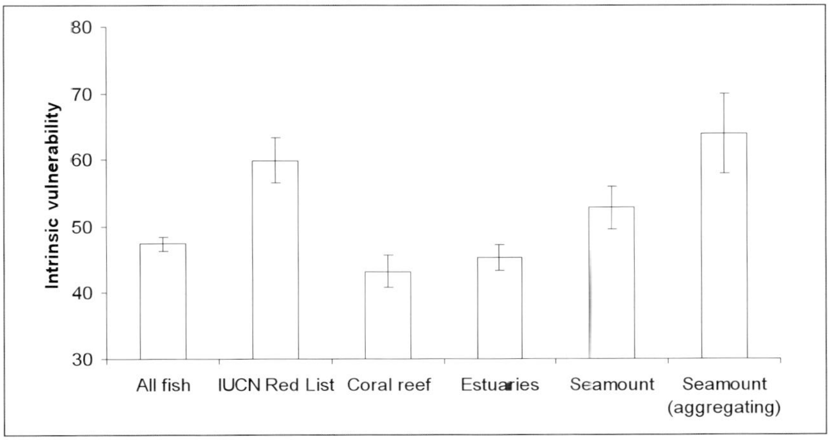


Coral reef-associated (N = 243), estuaries-associated (N = 381), seamount-associated (172), seamount-aggregating (N = 15), all fish (N = 1,353), species listed under the IUCN Red List (N = 161)
The error bars represent 95 per cent confidence limits (redrawn from Cheung *et al.*, 2007). Higher intrinsic vulnerability index represents higher vulnerability to fishing.

Fish living in coral reefs are often small-bodied, fast growing species which tend to have lower intrinsic vulnerability. However, because fish diversity in coral reefs is high, the number of species that are vulnerable to fishing is also high (Cheung *et al.*, 2007).

## 3. Depletion of vulnerable fish

The rapid increase in fishing intensity in the last six decades has already depleted the most vulnerable fish stocks. As a result, the ocean has become increasingly dominated by small-bodied, fast-growing, less vulnerable species (Figure 8). We compared the composition of fish in the sea with different intrinsic vulnerability in each year from 1950 to 2004. The composition of fish stocks is indicated by the contribution of each fish stock to the total catch, or in other words, the average intrinsic vulnerability of the fish weighted by their annual catch (Cheung *et al.*, 2007). The results show that average intrinsic vulnerability of fish in each year's global catch has decreased steadily and significantly since the 1950s (Figure 8a). When we excluded the highly productive and variable small pelagic fish (i.e. pelagic fish less than 30 cm in total length) from our analysis, the decline in average intrinsic vulnerability of the annual catch became even steeper (Figure 8b). The rate of decline increases further when we focus on demersal fish only (Figure 8c). A similar trend is observed for fishes caught by the member nations of the Commonwealth (Figure 8d).

### Figure 8: Mean intrinsic vulnerability index of catch from 1950 to 2004

(a) All fish taxa

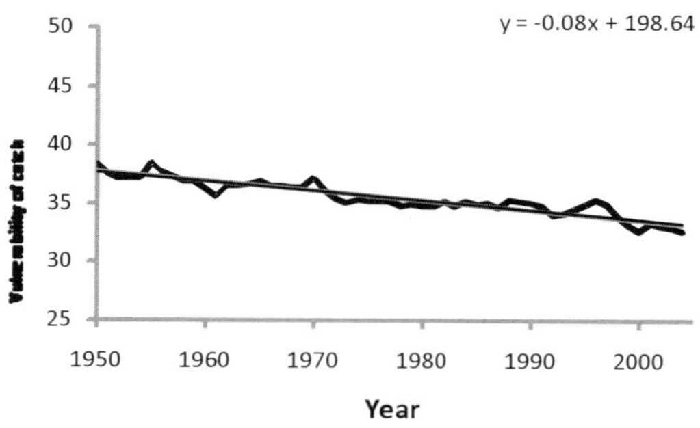


(b) All fish taxa except small pelagic fish (< 30 cm in total length)

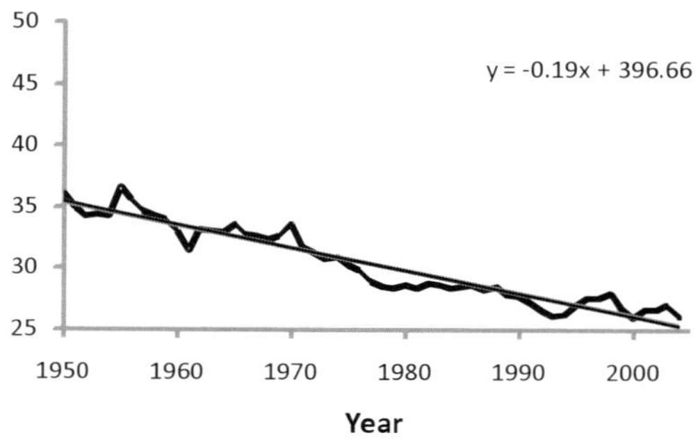


(c) Demersal fish species only

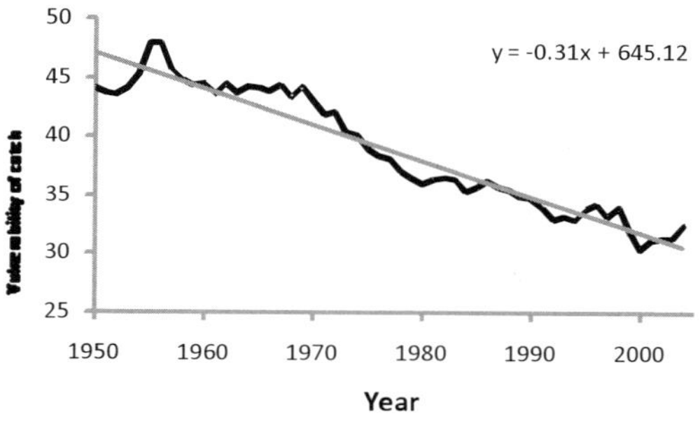


(d) All fish taxa caught from EEZs of Commonwealth countries

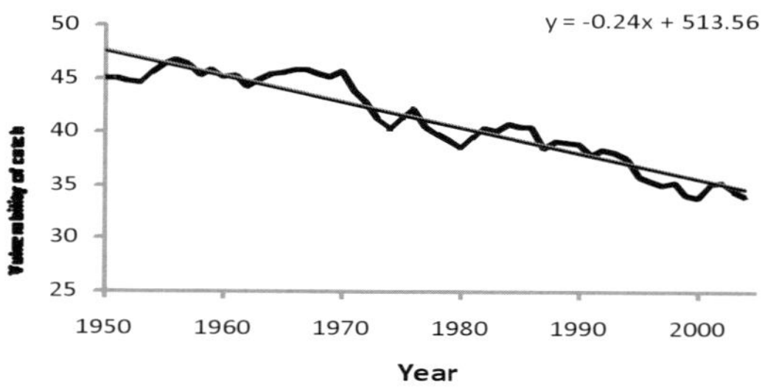


Note: We included taxa that are reported as family-, genera- or species-level groups in the catch statistics only.

In addition to a structural change in fish communities over time, depletion of vulnerable species has also expanded from the continental shelf to the deep seas and high seas. We mapped the average intrinsic vulnerability of annual fish catch in the 1950s, 1970s and 2000s onto a 30' lat. X 30' long. grid of the world ocean (Figure 9). In the 1950s, fisheries were targeting the abundant large-bodied, vulnerable fish stocks along the continental shelf (Figure 8a). By the 1970s, catch in these areas was increasingly dominated by less vulnerable fish because the more vulnerable stocks were depleted. Simultaneously, catch of vulnerable fish stocks were increasingly concentrated in the high seas and deep seas. By the 2000s, catch from most of the world ocean became dominated by less vulnerable species, except those from the mid-oceanic ridges in the North Atlantic and seamounts in the South Pacific, where catch is still dominated by highly vulnerable fish stocks. The deepening of fishing activities is also clear from the significantly increasing catch of deeper water species since the 1950s (Morato *et al.*, 2006), that are partly supported by subsidies provided by countries to the fishing sector (Sumaila *et al.*, 2006).

**Figure 9: Global maps of mean intrinsic vulnerability of exploited taxa weighted by their catch**

(a) 1950s

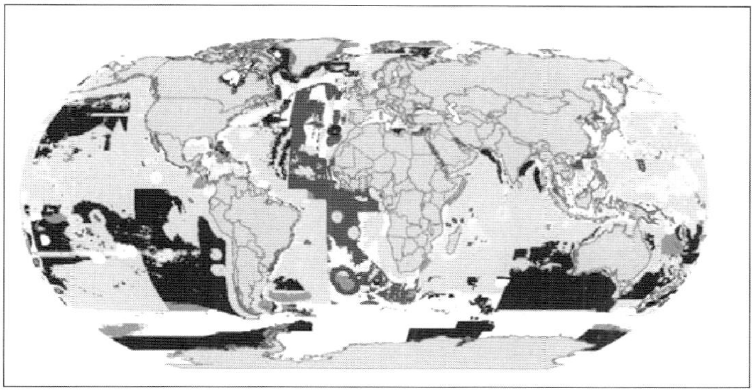

(b) 1970s

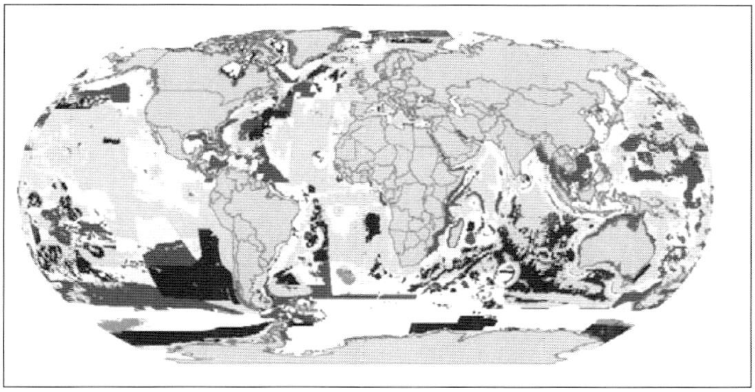

(c) 2000s

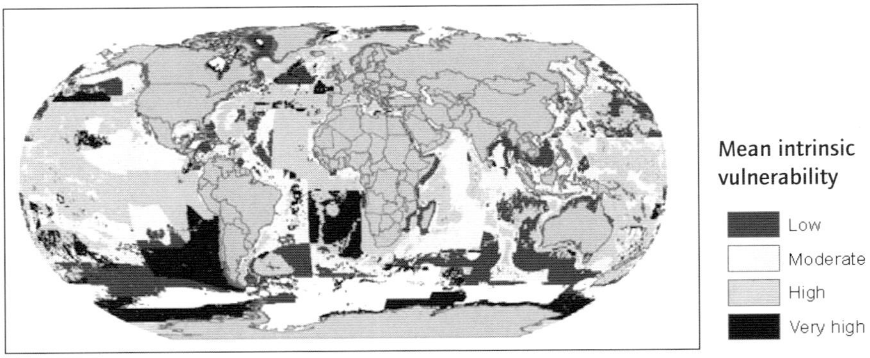


## Discussion

Following the huge increase of global fishing effort since the 1950s, global exploited fish stocks have reached their maximum sustainable limit in the last few decades. In ocean basins where industrial scale fishing occurred in the 19th or early 20th century, maximum limits were reached as early as the 1970s. The increasing global catch was made possible only by the expansion of fisheries to new fishing grounds in the continental shelf. Advanced technology also allowed fishing to expand to deeper waters and to previously inaccessible seamount resources (such as those in seamounts). However, it is clear from the results presented in this report that no unexploited fishing grounds now remains. With the over-exploitation and depletion of fish stocks, if effective management is not in place, the decline in global fish catch is likely to continue.

The large expansion of fishing effort and fishing grounds has altered the structure of marine ecosystems. Mean trophic level of catch, an indicator of fisheries impacts on marine ecosystem, declined considerably relative to the 1950s (Pauly *et al.*, 1998). Trophic level is a measure of the position of an organism in the food chain. It scales from 1 to 4.5 in most marine ecosystems, with animals higher in the food chain having higher trophic levels. As large predatory species are depleted by fishing, marine ecosystems, and the catch extracted from these systems, are increasingly dominated by lower trophic level species. This is happening in most parts of the world oceans, resulting in a decline in mean trophic level of catch (Pauly *et al.*, 1998). For example, in the North Atlantic, biomass of predatory fishes with trophic level of 3.75 or more declined by two thirds over the past five decades (Christensen *et al.*, 2003). This is similar to the large-scale removal of large animals in terrestrial systems by human activities and the increased dominance of the less vulnerable species, or the removal of vulnerable primary forest. In fact, the unprecedented large-scale extraction of fish from the ocean in the history of humans has been suggested to cause observable and evolutionary changes in the biological characteristics of exploited organism (e.g. Law, 2000; Conover and Munch, 2002; Darimont *et al.*, 2009). This may further reduce population productivity and stability, and affect other organisms that are dependent on the exploited species (Darimont *et al.*, 2009). Besides, the increasingly dominant small-bodied, short-lived, lower trophic level fishes stocks are more strongly affected by environmental fluctuation and changes. Particularly, climate change is likely to have considerable effects on marine ecosystems (Cheung *et al.*, 2009). Fisheries that are dependent on these species may suffer from the high variability of these fish stocks.

Some fish communities are particularly vulnerable to exploitation, notably seamount fish communities and deepwater species. Advanced fishing technology is allowing commercial fishing in deep waters and on seamount fisheries resources, but these fisheries are often boom-and-bust in nature (Morato *et al.*, 2006; Cheung *et al.*, 2007). Fisheries tend to deplete a patch of fisheries resources before moving to another fishing ground (Frekets *et al.*) and, as a result, fish stocks are quickly depleted. Given past experience, it is very likely that the current pattern of deepwater and seamount fisheries could not be sustained for long.

Some fishing activities not only affect fish stocks but also destroy their habitats. Fishing by bottom trawlers has increased dramatically since the 1950s, and this practice destroys biogenic bottom structures that are important for the survival of adult and juvenile fish as well as other marine organisms (Watling and Norse, 1998). This has resulted in shifts in ecosystems on the continental shelf of most of the world's ocean, with depletion of intrinsically vulnerable demersal fishes and an increase in less vulnerable small pelagic fishes.

We have, in this chapter, shown that the pattern of over-exploitation and vulnerability of fish stocks at a global scale reflects the situation at regional scales too. In fact, the changes observed in the waters of the member states of the Commonwealth are very similar to the global pattern. Many fish stocks in the Commonwealth share similar threats to those in other parts of the world, and therefore need urgent action to reverse the negative trends.

## Conclusion

The available evidence indicates quite clearly that global fisheries in general are not currently sustainable. Many intrinsically vulnerable fish stocks are over-exploited or depleted. Destructive fishing activities cause long-term damage to fish habitats. These result in shifts in ecosystem and fish community structures. The disturbing results presented in this chapter do not take into account the potential effects of climate change, which, according to recent papers (Allison *et al.*, 2009; Cheung *et al.*, 2009; Edward *et al.*, 2009), will have a disproportionately devastating impact on the ecology and fisheries of tropical developing countries, many of which are member states of the Commonwealth. There are many direct and indirect reasons leading to the unsustainability of our fisheries. One major factor is overcapacity of world fishing fleets, which are supported by government subsidies (Sumaila *et al.*, 2007). The problem is ubiquitous, and members of the Commonwealth are facing similar challenges. Particularly, all members of the Commonwealth share a part of the world ocean, many of them have long history and culture of fishing and other related industries. It is undesirable to see these traditions destroyed because of ineffective fisheries management and unsustainable fishing practices. There is still hope that we can continue such traditions, providing food supply and incomes to coastal communities while having a healthy marine environment. With the depletion of biomass in the world ocean, it is clear that fisheries management should be steered towards rebuilding of marine ecosystems and protection of vulnerable fish stocks that cannot sustain the current level of fishing (Pitcher *et al.*, 1998; Pauly *et al.*, 2002).

## References

Bailey, M., Rotinsulu, C. and Sumaila, U. R. (2008) The Migrant Anchovy Fishery in Kabui Bay, Raja Ampat, Indonesia: Catch, Profitability, and Income Distribution, *Marine Policy*, 32, pp. 483–488.

Berkes, F., Hughes, T. P., Steneck, R. S., Wilson, J. A., Bellwood, D. R., Crona, B., Folke, C., Gunderson, L. H., Leslie, H. M., Norberg, J., Nyström, M., Olsson, P., Österblom, H., Scheffer, M., and Worm, B. (2006) Globalization, roving bandits, and marine resources, *Science*, 311(5767), pp. 1557–1558.

Cheung, W.W.L., Lam, V.W.Y., Sarmiento, J.L., Kearney, K., Watson, R. and Pauly, D. (2009) Projecting global marine biodiversity impacts under climate change scenarios, *Fish and Fisheries* 10, pp. 235–251.

Cheung, W.W.L., Pitcher, T.J. and Pauly, D. (2005) A Fuzzy logic expert system to estimate intrinsic extinction vulnerabilities of marine fishes to fishing, *Biological Conservation*, 124, pp. 97–111.

Cheung, W.W.L., Watson, R., Morato, T., Pitcher, T.J. and Pauly, D. (2007) Intrinsic vulnerability in the global fish catch, *Marine Ecology Progress Series*, 333, pp. 1–12.

Christensen, V., Guénetter, S., Heymans, J. J., Walters, C. J., Watson, R., Zeller, D. and Pauly, D. (2003) Hundred-year decline of North Atlantic predatory fishes, *Fish and Fisheries* 4, pp. 1–24.

Conover, D.D. and Munch, S.B. (2000) Sustaining fisheries yields over evolutionary time scales. *Science* 297, pp. 94–96.

Darimont, C.T., Carlson, S.M., Kinnison, M.T., Paquet, P.C., Reimchen, T.E. and Wilmers, C.C. (2009) Human predators outpace other agents of trait change in the wild, *Proceedings of The National Academy of Sciences*, 106, pp. 952–954.

Edward, H. A., Perry, A. L., Badjeck, M., Adger, W. N., Brown, K., Conway, D., Halls, A. S., Pilling, G. M., Reynolds, J. D., Andrew, N. L., Dulvy, N. K. (2009) Vulnerability of national economies to the impacts of climate change on fisheries, *Fish and Fisheries*. 10.1111/j.1467–2979.2008.00310.x.

FAO (2006) *The Status of World Fisheries and Aquaculture* (Rome: FAO).

Hutchings, J. A. and Reynolds, J. D. (2004) Marine fish population collapses: consequences for recovery and extinction risk, *BioScience*, 54(4), pp. 297–309.

Law, R. (2000) Fishing, selection and phenotypic evolution, *ICES Journal of Marine Science* 57(3), pp. 659–668.

Morato, T., Watson, R., Pitcher, T. J., Pauly, D. (2006) Fishing down the deep. *Fish and Fisheries*, 7, pp. 24–34.

Myers, R. A., and Worm, B. (2003) Rapid worldwide depletion of predatory fish communities, *Nature*, 432, pp. 280–283.

Pauly, D., Christensen, V., Dalsgaard, J., Froese, R., and Torres, F. Jr. (1998) Fishing down marine food webs, *Science*, 279, 860–836.

Pauly, D., Christensen, V., Guénette, S., Pitcher, T. J., Sumaila, U. R., Walters, C. J., Watson, R. and Zeller, D. (2002) Towards sustainability in world fisheries, *Nature* 418, pp. 689–695.

Pitcher, T. J. and Pauly, D. (1998) Rebuilding ecosystems, not sustainability, as the proper goal of fishery management. *Reinventing Fisheries Management* (London: Kluwer Academic Publishers), pp. 311–329.

Reynolds, J. D., Dulvy, N. K., Goodwin, N. B. and Hutchings, J. A. (2005) Biology of extinction risk in marine fishes, *Proceedings of the Royal Society B*, 272, pp. 2337–2344.

Sumaila, U. R., Khan, A., Teh, L., Watson, R., Tyedmers, P. and Pauly, D. (2006) Subsidies to high seas bottom trawl fleet and the sustainability of deep sea benthic fish stocks. *Catching more bait: A bottom-up re-estimation of global fisheries subsidies*. Fisheries Centre Research Reports 14(6) (Vancouver: Fisheries Centre, University of British Columbia) pp. 47–51.

Sumaila, U. R., Khan, A., Watson, R., Munro, G., Zeller, D., Baron, N. and Pauly, D. (2007) The World Trade Organization and global fisheries sustainability, *Fisheries Research*, 88, pp. 1–4.

Watling, L. and Norse, E. A. (1998) Disturbance of the seabed by mobile fishing gear: a comparison to forest clearcutting, *Conservation Biology*, 12, pp. 1180–1197.

Watson, R. and Pauly, D. (2001) Systematic distortions in world fisheries catch trends, *Nature* 414, pp. 534–536.

Watson, R., Kitchingman, A., Gelchu, A. and Pauly, D. (2004) Mapping global fisheries: sharpening our focus, *Fish and Fisheries*, 5, pp. 168–177.

World Bank and FAO (2008) *The Sunken Bilions. The Economic Justification for Fisheries Reform.* Agriculture and Rural Development Department (Washington DC: The World Bank).

Worm, B., Barbier, E. B., Beaumont, N., Duffy, J. E., Folke, C., Halpern, B. S., Jackson, J. B. C., Lotze, H. K., Micheli, F., Palumbi, S. R., Sala, E., Selkoe, K. A., Stachowicz, J. J., Watson, R. (2006) Impacts of biodiversity loss on ocean ecosystem services, *Science*, 314, pp. 787–790.

# 15 Reversing Perverse Incentives: The Case for a Rights-based Strategy

G.T. (Stan) Crothers and James Wilen

## Introduction

This chapter discusses the common property/common pool attributes of fisheries and notes that all of the most successful fisheries management regimes, to a greater of lesser extent, adopt a rights-based approach.

It observes the current state of the world's fisheries and identifies common themes. It looks at the consequent challenges facing Commonwealth Governments seeking to manage their fisheries in successful ways that ensure maximum benefits are produced, whilst ensuring biological sustainability. It discusses these issues in relation to near shore fisheries, off shore fisheries and high seas and highly migratory fisheries. It explicitly highlights the importance of accountable and cost effective governance and management arrangements.

## Global Fisheries Management: An Overview

A canvas of scientific and popular literature on the state of the world's fisheries leaves a very gloomy impression (Sumaila and Cheung, this volume). The most highly-cited papers project long term trends that include declining overall harvest, serial depletion of top predators, the replacement of upper trophic level species by 'jellyfish and slime', and even complete extinction of all species within 50 years (Pauly *et al.*, 1998, Myers and Worm, 2003, Worm *et al.*, 2006, Christensen *et al.*, 2003). Much of the literature reports and repeats these gloomy descriptions without offering concrete solutions. Other papers offer a range of ideas without much sense of consensus about how to fix the system or even prognostications about whether a fix is possible at this stage (Pauly *et al.*, 2002; Pitcher and Pauly, 1998). It is difficult in the face of such dire predictions to have much hope that the world's fisheries can ever be harvested sustainably and beneficially.

But in fact, we have learned an enormous amount about fisheries over the past 50 years. We do understand why some of the trends we have witnessed have occurred, and we also understand how to fix the adverse outcomes. Resource economists have known for 50 years that the cause of overexploitation is the perverse incentives driven by legal and political institutions within which fisheries are prosecuted (Gordon, 1954). In particular, resources that are owned by no one, and that are utilised as open access commons, generate perverse incentives to overharvest. When resources are common property, no owner or entity controls the total amount of effort employed and the harvest taken. Rather than a single owner/manager making coordinated decisions about how many boats to use and how many fish to take, many individuals make atomistic decisions in response to their own incentives and payoffs. These individual decisions ignore the impact of the individual on the system. The important point is that fishermen are not evil or inherently anti-conservation; rather they are compelled by the legal institutions within which they operate to make decisions that reflect perverse incentives.

The perverse incentives that are generated by the absence of ownership and control have two important consequences, one on the resource and the other on users. The impact on the resource is almost always to degrade its productivity, so that biomass levels are driven to low levels, growth rates are reduced and population structure is altered towards smaller fish. In extreme cases, resources can be driven to small fractions of original natural levels, levels at which productivity is also a small fraction of potential maximum sustainable yields. The impact on resource users is that economic returns are driven down to the minimum levels necessary to sustain users. Thus there is no surplus generated, and the potential economic surplus that might be generated is wasted by too many users. Open access over-exploitation thus makes the pie smaller overall, and divides that pie into pieces that are individually very small. Generally, an overexploited resource can yield the same harvest or even higher levels with less effort than will be attracted under open access commons use. Thus it is possible in principle to make the pie larger, and give each existing participant a bigger slice, if the proper policies can be adopted.

## Fisheries since UNCLOS

Prior to 1978, most of the world's fisheries operated under open access commons conditions. The institutional structure within which the world's fisheries were managed changed dramatically with the United Nations Convention on the Law of the Sea (UNCLOS) negotiations and the subsequent extension of jurisdiction in 1978. By extending jurisdiction to 200 miles and granting control of formerly common property resources to coastal nations, the Law of the Sea Treaty provided a foundation for sustainable and profitable management of those resources (Barnes, this volume). The extension of jurisdiction was only a partial fix and it did not, by itself, solve the overharvesting and fisheries dissipation problem. While the extension of jurisdiction to 200 miles enclosed about 90 percent of the world's fisheries, a number of important fisheries remained outside the jurisdiction and hence still subject to the perverse incentives generated by open access use. Many large migratory species are still vulnerable and endangered because they are still subject to perverse incentives associated with the high seas common pool incentives.

## Failures and Successes

The most prevalent failures in fisheries management have been associated with the lack of appreciation of the institutional cause of perverse behavioural incentives. In fact, much fisheries regulation and policy is aimed at the symptoms of open access resource use, rather than the cause (Wilen, 2006). For example, excessive entry of fishermen under (domestic) open access conditions has been tackled most frequently by stifling the effects of entry on fishing mortality rather than by tackling the reasons for excessive entry. Many fisheries are managed by reducing the season length or restricting areas fished or by reducing gear effectiveness as effort grows. But reducing season length or gear effectiveness does nothing to stem the tendency for effort to enter under open access. In fact, reducing season length often induces a 'race to fish' that typically involves individual fishermen adding more capital and fishing power in order to access fish before their peers do, and before regulators close the season in response to growing effort.

The most successful cases of fisheries management over the past three decades have eliminated the race to fish by recognizing that the focus needs to be on the cause of the problem rather than symptoms. In particular, successful management involves altering legal institutions within which individual fishermen make everyday decisions. Most importantly, fishermen need to move from an institutional framework in which they are competing for an unknown and uncertain share of potential harvest, to one where they have secure rights to a certain share of potential harvest. Under the race to fish, fishermen compete wastefully with each other to increase their share of the potential harvest before regulators close the fishery. In fisheries where fishermen have a right to a share of potential harvest, they do not need to compete with each other for fish landed; instead they can turn their efforts and ingenuity toward increasing the value of their secure share.

## Rights-based Management

The most important feature of modern success stories in fisheries management is that they have reoriented fundamental behavioural incentives away from wasteful competition toward more constructive value creation. This has been achieved with several different institutions, each an example of a 'rights-based' scheme. There are basically three broad types of rights-based systems.

- **Individual Transferable Quota schemes (ITQs).** ITQs grant fishermen rights to harvest shares of a biologically determined allowable catch each season. Fishermen thus are guaranteed rights to land X tons of fish (which may change from year to year depending upon the health of the fishery), and those fish can be landed at any time in the season, with any allowable gear types, and marketed into any available niche.
- **Harvester Cooperatives.** Co-ops grant shares of an allowable quota not to individuals, but to groups. Co-ops must adopt additional actions within the cooperative to convert the group share allocation into a successfully functional entity. In particular, once a co-op has been granted rights to land certain quantities as a unit, decisions must be made about how to operationally manage the co-op share. In some cases, the co-op share is simply devolved to the member units as internal ITQs, and each fisherman acts independently to manage his/her own share of the co-op share. At the other end of the spectrum, a cooperative may be operated as a sole entity much like a corporation, with decisions made about all internal operations in ways that maximise overall returns, with distribution of proceeds by some prearranged formula. These kinds of co-ops have generated dramatic efficiencies by reducing redundant fishing effort, coordinating the use of

vessels over time and space, and altering product mix and output to take advantage of market conditions (Wilen and Richardson, 2008).

- **Territorial Use Right Fisheries (TURFs).** A third type of rights-based institution also grants rights to a group or cooperative, but instead of being species-based, this type of institution grants the group rights to use a unit of space in a marine ecosystem. These are common in island fisheries in the South Pacific as well as in developing and some developed countries. For TURFs to be most effective, it is necessary to create an internal management structure that manages the harvest of species within the allocated space. In practice, this again manifests itself across a spectrum, from more or less hands off systems that devolve use decisions to individuals, to others that allocate equivalents of internal ITQs to members, to still other systems that manage and coordinate all aspects of the group's effort, from the distribution of fishing effort over time and space, to the distribution of products and species types into local and international markets (Cancino *et al.*, 2007).

Rights-based systems have generated successful results across a number of conservation performance measures. In the final analysis, conservation success is always determined by the effectiveness of managers at determining and implementing sustainable allowable catches. This task is generally made easier under rights-based management. First, total allowable catches under ITQs are met precisely, rather than as indirect outcomes of attempts to control elements of fishing mortality. Problems associated with the race to fish such as ghost fishing from abandoned gear are reduced or eliminated. In addition, the reduction of effort and elimination of the race to fish removes the inherent tussle between fishermen who are trying to increase effort and regulators who are trying to stifle effort. Indeed, once rights holders are granted some security of tenure in these systems, they have enhanced incentives to ensure that safe and sustainable biological objectives are pursued and adopted. A significant advantage of rights-based systems is that they engage fishermen as stakeholders interested in long term productivity of the resource, rather than as adversaries placed in a continuous and escalating struggle with regulators to control fishing mortality (Wilen, 2006).

Aside from positive effects on conservation, rights-based systems reorient basic incentives away from a race to fish, towards a race to create individual and collective value and wealth. This is the most dramatically impacted feature of rights-based systems compared with all others. A point not well appreciated by fisheries scientists or fisheries managers is that open access common pool incentives guarantee that fishermen will almost always be on the edge of insolvency. A central prediction of the Gordon model is that entry will continue until every entrant is earning just enough to pay expenses. In an efficiently operated fishery, entrants would not only earn enough to pay expenses, but they would also earn enough to pay for the inherent biological productivity of the resource.

## Experience with Rights-based Management

The accumulating evidence from the several score of fisheries that have adopted rights-based management reveals a number of important summary observations. First, eliminating race to fish incentives immediately changes the character of the fishing process, away from a frantic and adversarial system always on the brink of insolvency, to a more leisurely and measured process that not only pays the (reduced) expenses, but generates a surplus to participants. Second, once fishermen are granted secure rights, the nature of their involvement with regulators and the reach of their horizon changes, from short-term adversarial, to long-term cooperative. This enables new opportunities for fishermen to absorb some of the costs and responsibilities of information gathering and management, and it unleashes new opportunities for regulators to take advantage of stakeholders who have new and common interest in sustainable and efficient management. Third, the 'race to value' that substitutes for the race to fish unleashes a torrent of innovation that serves largely to increase the values that everyone derives from the resource (Wilen, 2004). New (often slower) fishing methods are adopted that catch larger or higher-valued fish and keep them in better condition when delivered to processors. Vessels are consolidated and reconfigured in order to catch fish with fewer inputs and less expensively. New products are created; new markets are developed (Casey *et. al.*, 1995); investment in brand identification and cooperation in certification programs become feasible (Tindall, this volume). Overall, a new culture is created that aims ingenuity towards deriving more net value from resources that were previously generating zero net value.

Perhaps the most surprising outcome from the experience that is accumulating as we watch success stories unfold is how large the potential surplus actually is. Following Gordon's early work, fisheries economists once maintained the view that most rent dissipation and waste under common property was associated with too many boats. But case studies have revealed that rents are wasted across many dimensions, from too many vessels, to inefficiently configured inputs, to poor product quality and limited market development, to inefficient use of space, etc. The evidence from the implementation of rights-based fisheries is that the rent creation process 'unwinds' the rent dissipation process, presenting myriad opportunities for wealth generation across a multitude of decision margins. The creativity that is unleashed after fishermen's incentives are reoriented to maximise value is the wellspring of wealth generation in modern fisheries (Wilen, 2004).

A good gauge of the potential economic surplus that is being wasted in conventionally-managed fisheries is the lease value that emerges in rights-based fisheries. The lease value represents a measure of economic surplus in the same way that lease values for land represent the net proceeds over and above expenses that users are willing to pay for the productive services of land. Lease values in rights-based fisheries are surprisingly large, of the order of 50–80 percent of the ex-vessel value. In conventionally managed fisheries, costs rise to the level of revenues through excess entry, excess application of inputs, and general inefficiencies caused by the scramble for common pool fish. In rights-based fisheries, stakeholders reduce effort and consolidate and reconfigure to achieve cost savings. These cost savings are part of what emerges in the lease values, with a remainder associated with revenue increases from market innovation. If we use a back of the envelope total global value of marine fisheries of $100 billion per year, these numbers suggest that the amount of value that is being wasted by conventional fisheries institutions may be of the order of $50–80 billion per year (Wilen, 2004; World Bank, 2009). In comparison, the actual value created by the world's fisheries today is closer to zero, or negative (due to the need to subsidise).

## Fisheries Management in Developing Countries

The developing countries pose special problems for effective redesign of fisheries management institutions. Many countries have potentially productive resources within their EEZs that are in various states of over-exploitation and misuse. But they lack the supporting institutional infrastructure that may be important to redesigning and implementing new fisheries institutions. It is convenient to divide fisheries in developing countries into offshore and inshore fisheries. Offshore fisheries are mostly industrial fisheries, exploited by large and capitalised vessels, and pursuing a select group of species. Vessels may be foreign or domestic fleets fishing illegally, or fishing as concessionaires. The terms of use may be transparent, with biological oversight and lease/rent capture for the coastal state, or under opaque and potentially corrupt side deals without concern for long term sustainability. Nearshore fisheries are often prosecuted by large numbers of small vessel fishermen, pursuing many species for mostly local markets, and with virtually no biological oversight or attempt at sustainable management. Nearshore fisheries generate little economic surplus, but they contribute in important ways to diet and nutrition, and to employment in areas with few opportunities.

## Developing Countries: Offshore Fisheries

Developing countries generally suffer capital scarcity and as a result may not have domestic fleets capable of utilising offshore fisheries. As a result, offshore fisheries in many developing countries are prosecuted by foreign fleets, either illegally or legally. In situations where foreign fleets fish illegally, resources return no rents to adjacent coastal states and the resource is likely to be overexploited. When foreign fleets fish legally under concession arrangements, the rents that are transferred to the coastal nation depend upon agreements, which may be negotiated transparently and may benefit the population broadly. Such agreements may also be done corruptly, with behind the scenes rent transfers that do not benefit the citizens of the coastal state directly (Standing, 2008, cited in Davies and Bergh, this volume). Because coastal nations in developing countries lack infrastructure to manage their fisheries, decisions about allowable catches are likely to be *ad hoc* and not necessarily reflecting best available scientific information about sustainability.

Offshore fisheries in developing countries may be candidates for rights-based systems that are similar to successful examples in developed countries. In principle, developed country governments could auction or grant licenses to harvest certain quantities, and capture rents from the production of offshore marine resources. In practice, such schemes require legal infrastructure, human capital necessary to design and implement such schemes, and biological expertise sufficient to conduct stock assessments and determine allowable catches. Sustainable use of offshore resources also requires enforcement infrastructure to police offshore activities, to monitor adherence to contractual arrangements, and to arrest and prosecute those engaged in illegal activity. In addition, if fisheries are to benefit the broad populace of developing countries, rent capture and its distribution must be made transparently without corruption and capture by powerful elites. Certainly these preconditions are significant hurdles, but it is also true that the potential gains can be significant (Cunningham and Nieland, 2005; Cunningham *et al.*, in press).

## Developing Countries: Nearshore Fisheries

Nearshore fisheries in developing countries present different and more complicated issues. Because there are typically many small-scale fishermen involved, fishing in small vessels in pursuit of numerous species, the task of implementing any top-down rights-based and species-based institutions like ITQs seems daunting. A more plausible model may be a hybrid of group cooperative rights coupled with TURFs. These would require carving up coastal fisheries into well-defined physical zones, and granting exclusive rights to resources within those zones to groups such as local harvester cooperatives. The advantages to this kind of scheme are that it creates incentives among stakeholders to manage with bottom-up and locally-based knowledge, to implement and enforce rules that are internally acceptable, and to self-police against illegal poaching by non-members. Such a system requires identifying a closed class of stakeholders and legitimising their exclusive rights to reap the rewards from responsible self-management. In other words, it requires that the federal government legitimise and legally sanction the exclusivity of the group rights in the first place, as well as the particular spatial delineation of the TURFs. This task will be difficult in regions with cultural histories that include open access to coastal resources, but it is a necessary precondition to managing fisheries conservatively and that generate real wealth.

Place-based group rights systems have been implemented and have operated for centuries in the nearshore fisheries off Japan's coast (Uchida and Wilen, 2004). These systems grant rights to members of harvester cooperatives to utilise and manage resources within specifically designated zones off coastal communities. Chile has also adopted such systems along the lengthy Chilean coastline. In both countries, fisher co-ops manage their own marine systems, with some advice and oversight from scientists at the national levels. Both systems have generated a spectrum of outcomes, since each TURF has been left to evolve its own system of self-management. Some co-ops have not evolved much beyond business-as-usual operation by the original grantees, including rent dissipation and resource overuse. But the vast majority of cases have witnessed improvement in both the health and sustainability of the resources, as well as improvements in profitability and efficiency. Some have implemented dramatic and highly coordinated internal management systems that allocate effort over time and space to maximise yields and that manage harvest over time to take advantage of favourable market conditions (Cancino *et al.*, 2007).

It is clear that the success of place-based cooperatives depends upon many factors, including the biological productivity of the space, the number and cohesiveness of fishermen who become stakeholders and the quality of leadership, education and knowledge. But the critical point is that these systems convert open access common pool fisheries into entities that have the capability of redirecting individual incentives to activities that maximise group returns.

## Management of High Seas and Highly Migratory Species

As discussed earlier, the central focus of the current legal and policy framework that governs the conservation and management of fisheries is UNCLOS. It is UNCLOS that defines the rights that allow coastal states to govern and manage the harvesting of fish within their Exclusive Economic Zone (EEZ) and the obligations they must meet when doing so.

A significant proportion of the world's fisheries are within EEZs and UNCLOS provides nation states with the mandate and the means to effectively manage their domestic fisheries in a way that ensures both positive environmental and economic outcomes (Barnes, this volume) The performance of individual nation states in managing their domestic fisheries has to date been mixed in relation to fisheries sustainability and dismal in relation to wealth generation (FAO, 2009; World Bank, 2009; the overview at the beginning of this chapter).

UNCLOS is also the instrument that confirms that all states have freedom to fish on the high seas and defines the obligations attached to that right. The potential for high seas and highly migratory species to generate significant sustainable wealth means that it is important to continue to focus on the challenge of effectively managing those fisheries that UNCLOS defines as outside the exclusive authority of any one nation state.

As with near shore and off shore fisheries, the primary challenge for nation states in relation to high seas and highly migratory species is how to ensure both the biological sustainability of fisheries and that the maximum level of wealth can be generated by lawful fishing activity. This challenge is especially difficult to address given the common pool characteristics and current common property attributes of high seas and highly migratory fisheries. As evidenced in the Fisheries Management Overview of this chapter the undisputed characteristic of open access fisheries is that they will inevitably be biologically overfished and wealth will be dissipated.

Experience established that UNCLOS alone did not provide the capacity to effectively manage high seas (see Barnes, this volume). Continued depletion of fish stocks after UNCLOS (Alaskan Pollock and Atlantic cod fisheries) led to enhancement of the UNCLOS regime by adoption of the 1995 United Nations Fish Stock Agreement.

This agreement has allowed some meaningful progress to be made in managing fisheries sustainably, although there continues to be management failures (eg northern and southern blue fin tuna fisheries). The enhanced framework has not, however, created effective tools able to address the common property/pool characteristics of these fisheries. Specifically, it has not created mechanisms that:

- Enable accountable governance and management mechanisms to evolve;
- Create and distribute sustainable wealth generated by fishing activity;
- Result in incentives to ensure appropriate levels and type of investment in fisheries management (Crothers and Nelson, 2006).

There are numerous examples available that demonstrate the inability of existing institutional arrangements to achieve positive management outcomes for fisheries. (Haughton, this volume). It is clear that there is no lack of will on the part of governments to ensure sustainability for fish stocks and there is a demonstrated willingness to commit resources to fisheries management both domestically and internationally. Governments are investing in a multiplicity of fisheries management organisations but are not achieving desired fisheries management outcomes that are consistent with the expectations of their citizens or with their level of investment (High Seas Task Force, 2006).

The available evidence leads to a compelling conclusion. It is not the *quantity* of government action that results in less than optimal outcomes. It is the *quality* of outcomes that government action can achieve within the limitations of the current legal and policy frameworks that are inherently restricted. Current frameworks do not provide management organisations with effective management tools to achieve both biological sustainability and maximisation of wealth.

UNCLOS, together with its supporting international agreements such as the Fish Stocks Agreement, provides a framework for collective action by nation states that would allow them to manage high seas and migratory species to ensure the biological sustainability of the fish stocks. The open access framework for high seas that is generated by UNCLOS does not provide an effective way for nation states to exclude free rider nation states. Nor does it align incentives to maximise wealth.

The lack of management tools to influence all the factors that combine to ensure optimal outcomes means that individual nation states will continue to invest to maximise their own self interest – in the same way that individual fishers do in open access fisheries within a specific EEZ.

A two pronged strategy is necessary to allow nation states to move beyond the limitations of the current fisheries management frameworks.

First, maximise the performance of Regional Fisheries Management Organisations (RFMOs) within the existing frameworks created by UNCLOS and the Fish Stock Agreement (Haughton, this volume). UNCLOS does provide for a level of management that has a proven ability to improve on the outcomes achieved within open access fisheries. In combination with the Fish Stocks Agreement it is possible for nation states to manage for the biological well being of fish stocks and to manage the adverse impacts of fishing on the marine environment. Investing in the capacity and performance of RFMOs is essential to ensure there continues to be high seas fisheries from which wealth can be derived under a yet to be developed enhanced management framework.

Second, collectively invest in strategic policy and legal thinking to augment and improve current frameworks. UNCLOS arose from a collective awareness of the need to improve the tools available to nation states to manage the world's fisheries. The Fish Stock Agreement followed a collective commitment to improve the effectiveness of the tools available to nation states to achieve commonly agreed goals relating to fisheries. We are long past the need to make a further paradigm shift in the international policy and legal frameworks that mandate nation states to collectively manage fisheries. Only such an advance in approaches to fisheries management will allow nation states to fully capture the sustainable rents that high seas and migratory fisheries have the potential to generate.

The Commonwealth can play an important role in the first strategy by supporting member nations to participate effectively in RFMOs, and to build the capacity and capability of RFMOs relevant to member states. In particular, it can assist nation states with technical capacity to participate in RFMOs and facilitate the necessary levels of co-operation between nation states to improve the performance of RFMOs. The Commonwealth may also be able to make a valuable contribution by co-ordinating development aid for member states from within and outside of the Commonwealth to ensure that it is aligned with improving the capacity to RFMOs to preserve the status of key fish stocks.

It can contribute significantly to the second strategy by co-ordinating investment in improving the policy and legal frameworks and in helping governments to work together to build the consensus necessary to support a paradigm shift.

## Governance Arrangements

Sound governance and management arrangements are a pre-condition to achieving positive fisheries management outcomes. This requires governance and management arrangements that are accountable, effective and efficient. Specifying the roles and responsibilities of government(s), use rights holders and civil society is central to achieving this.

A number of the chapters (McClurg and Arbuckle, Tom'tavala, Bostock and Walmsley) in this publication identify the challenges that governance arrangements need to address in order to deliver positive fisheries management outcomes:

- Specifying objectives for the fisheries management regime in terms that maximise societal benefits;
- Specifying the environmental standards necessary to ensure the health of marine ecosystems and the sustainability of fish stocks;
- Specifying and allocating use rights, including protecting indigenous rights;
- Enabling use rights holders to collectively manage their harvesting activity consistent with the specified fisheries management objectives and environmental standards;

- Establishing incentives for sound investment in, and efficient delivery of, fisheries management services – such as administration, enforcement and research;
- Contributing to a fair trading system for sustainably harvested fish;
- Ensuring food security.

These chapters collectively establish that the greatest barrier to achieving enhanced fisheries management outcomes is the deficiencies of current fisheries governance and management arrangements. This is confirmed by the assessment of both the World Bank (2009) and FAO (2009) that world fisheries are, to a greater or lesser extent, in poor biological and economic health.

They equally establish that existing fisheries governance and management arrangement are not delivering the social, environmental, economic, and cultural outcomes that society expects. It is clear that these expectations will continue to grow and will place increasing pressure on fisheries management system that cannot within existing domestic and international policy and legal frameworks deliver on those expectations.

With increasing clarity over the last 30 years it has emerged that investing in further research or more enforcement delivered in the context of existing policy and legal frameworks will not relieve the pressure of society's legitimate expectations. It cannot now be disputed that the failure of the current framework is not a technical fisheries management issues. It is a political economy issue. To meet the legitimate expectations of society a paradigm shift is, therefore, required in policy and legal frameworks to make it possible to invest in and adopt sound fisheries governance and management arrangements. The Commonwealth can promote a systemic policy change which can help to sustain marine fish stocks for future generations.

UNCLOS does not itself constrain nation states from adopting a new paradigm for governance and management arrangements for domestic fisheries management. UNCLOS may constrain the ability to adopt a new paradigm for governance and management for high seas and highly migratory species. In real terms the constraint on adopting a paradigm shift is whether there is sufficient political will to build the consensus necessary to underpin a new paradigm. The environmental, economic and social pressures for change are real, compelling and of growing intensity. So too are the pressures to preserve the *status quo* – in particular, the vested self-interest of those who currently administer and benefit from the current arrangements.

This presents a significant opportunity for the Commonwealth and its agencies to support member states in building a consensus for change. It can also provide technical expertise in designing and implementing the new policy and legal frameworks and design and facilitate investment in the arrangements necessary to support the transition from the current to the new arrangements.

## References

Cancino, J., Uchida, H. and Wilen, J. (2007). TURFs and ITQs: coordinated versus decentralized decision making, *Marine Resource Economics*, 22(4), pp. 391–406.

Casey, K., C. Dewees, B.Turris, and J.Wilen. (1955) The effects of individual transferable harvest quotas in the British Columbia Halibut Fishery, *Marine Resource Economics*, 10(5).

Christensen, V., Guénetter, S., Heymans, J. J., Walters, C. J., Watson, R., Zeller, D. and Pauly, D. (2003) Hundred-year decline of North Atlantic predatory fishes, *Fish and Fisheries* 4, pp. 1–24.

Crothers G. T. (Stan) and Nelson, Lindie (2006) High Seas Fisheries Governance: A Framework for the Future, *Marine Resource Economics* 21(4).

Cunningham, S. and Neiland, A.E. (2005) *The role of fisheries in economic growth and poverty alleviation: towards a wealth-based approach to fisheries management*. Key-sheet prepared for the UK Department for International Development (DFID).

Cunningham, S., Neiland, A.E., Arbuckle, M. and Bostock, T. (in press) Wealth-based fisheries management: using fisheries wealth to orchestrate sound fisheries policy in practice. *Marine Resource Economics*.

FAO (2009) *The State of World Fisheries and Aquaculture 2008*. (Rome: FAO)

Gordon, H.S. (1954) The economic theory of a common property resource: the fishery. *Journal of Political Economy* 62, pp. 124–42.

High Seas Task Force (2006) *Closing the Net: Stopping Illegal Fishing on the High Seas*. Available at *http://www.high-seas.org*

Myers, R. A., and Worm, B. (2003) Rapid worldwide depletion of predatory fish communities, *Nature*, 432, pp. 280–283.

Pauly, D., Christensen, V., Dalsgaard, J., Froese, R., and Torres, F. Jr. (1998) Fishing down marine food webs, *Science*, 279, 860–836.

Pauly, D., Christensen, V., Guénette, S., Pitcher, T. J., Sumaila, U. R., Walters, C. J., Watson, R. and Zeller, D. (2002) Towards sustainability in world fisheries, *Nature* 418, pp. 689–695.

Pitcher, T. J. and Pauly, D. (1998) Rebuilding ecosystems, not sustainability, as the proper goal of fishery management. *Reinventing Fisheries Management* (London: Kluwer Academic Publishers), pp. 311–329.

Standing, A. (2008). *Corruption and industrial fishing in Africa*. U4 Anti-Corruption Resource Centre, p. 29, U4 issue 2008:7.

Uchida, H. and J. Wilen. (2004) Japanese coastal fisheries management and institutional design: a descriptive analysis, *Proceedings: 12th Biennial Conference of the International Institute of Fisheries Economics and Trade*, July 20–30, Tokyo, Japan.

Wilen, J. (2004). Property rights and the texture of rents in fisheries, in *Evolving Property Rights in Marine Fisheries* (D. Leal ed.) Rowman and Littlefield Publishers, Inc.

Wilen, J. (2006). Why fisheries management fails: treating symptoms rather than causes, *Bulletin of Marine Science*, 78, pp. 529–546.

Wilen, J. and F. Homans. (1997). Unraveling rent losses in modern fisheries: production, market, or regulatory inefficiencies?, in *Fisheries Management: Global Trends* (Huppert, D., Pitkitch, E., and M. Sissenwine, eds.) University of Washington Press.

Wilen, J. and E. Richardson. (2008). The Pollock Conservation Cooperative in *Case Studies in Fisheries Self-Governance* (Townsend, R., R. Shotton and H. Uchida, eds.) FAO Fisheries Technical Paper #504, FAO.

World Bank (2009) *The Sunken Billions: The Economic Justification for Fisheries Reform*. World Bank Group, Food and Agriculture Organisation, World Bank Publications, 2009.

Worm, B., Barbier, E. B., Beaumont, N., Duffy, J. E., Folke, C., Halpern, B. S., Jackson, J. B. C., Lotze, H. K., Micheli, F., Palumbi, S. R., Sala, E., Selkoe, K. A., Stachowicz, J. J., Watson, R. (2006) Impacts of biodiversity loss on ocean ecosystem services, *Science*, 314, pp. 787–790.

# 16 Marine Capture Fisheries in Crisis: A Commonwealth Call to Action

Richard Bourne and Mark Collins

## Introduction

With rapid growth of the world's human population, and the advanced technology now available to ever-expanding marine fishing fleets, the last half century saw an end to the idea that marine capture fisheries are an inexhaustible common good.[1] Clearly they are not (Sumaila and Cheung, this volume). The long term viability of fish stocks requires careful management, utilising a range of approaches, and mindful of the impact of climate change.

Catches from oceans and seas have remained relatively steady at 80–86 million tonnes since the mid-1990s, despite ever greater fishing effort. The UN Food and Agriculture Organisation (FAO) has warned that 75% of all marine fisheries are being exploited at or beyond their Maximum Sustainable Yield (MSY). Without better management a severe downturn in marine productivity and capture seems likely (Sumaila and Cheung, this volume), but by the time it is fully recognised it may be too late. Like the cod fisheries on the Grand Banks of Newfoundland, more fish stocks may become irretrievably exhausted unless urgent and concerted international effort is expended to address the problem.

Very recently an international group of fisheries experts have called for MSY to be re-interpreted as an absolute upper limit rather than as a target quota. There is evidence that fishing below this level can yield just as much fish but generate major ecological benefits at the same time (Worm, Hilborn *et al.*, 2009). They and others have also shown how management systems customised to the place and the people can contribute to the rebuilding of sustainable fisheries, for example in Kenya and New Zealand, and adaptable to a changing climate. There is evidence that timely reform of fisheries governance and control of the fishing effort, whilst painful in the short term, need not reduce the overall economic value of fisheries. To the contrary, such reforms will avoid over-use and destruction of fisheries resources (Barnes, this volume), while at the same time delivering greater economic value than is currently being obtained (World Bank and FAO, 2009).

The Commonwealth and its member states, for reasons explained below, need to be at the forefront of this agenda for change. There are five fundamental types of action indicated by the expert chapters presented in this book:

- Immediate application, where appropriate, of proven management solutions, such as lower catch quotas and rights-based community management, combined with economic incentives, zoning, and control of fishing gear, can deliver improved and sustainable economic, social and environmental outcomes (e.g. Crothers and Wilen; McClurg and Arbuckle, this volume);

- Replacement or removal of those legal, institutional and subsidy frameworks that exacerbate the fisheries crisis by creating perverse incentives to overfish (e.g. Von Moltke, this volume);
- Enhancement and enforcement of international controls on illegal, unreported and unregulated fishing (IUU) (e.g. Davies and Bergh, this volume);
- Appropriate measures to protect and restore the critical habitats and ecosystems on which coastal fisheries depend (e.g. Lang and Trebilco, this volume);
- Support for consumer-led sustainability programmes (e.g. Tindall, this volume).

## Fisheries and Development

The Commonwealth is a unique association of large and small states, embracing a wide spectrum of development needs. Achieving the Millennium Development Goals remains a key priority for all 53 countries, in particular to address the challenge of poverty and its associated impacts. If fisheries governance can be improved, considerable development benefits will accrue, from both industrial and artisanal fisheries.

Whereas the 1994 UN Convention on the Law of the Sea (UNCLOS) served to reduce open access by giving states sovereignty and sovereign rights over a proportion of the world's oceans, the management systems that emerged as a result have been less than effective, both in terms of preventing overfishing and in terms of unlocking the inherent wealth of fisheries on a sustainable basis (Barnes, this volume). In all regions of the world there is a trend of fish stock collapse. Whilst there are indications that locally appropriate management solutions can be effective in some regions (Haughton; Hanich and Tsamenyi; Tom'tavala; McClurg and Arbuckle; Crothers and Wilen, this volume), there is also evidence that some excess fishing effort in the developed world has simply been transferred to the developing world, where laws and their enforcement are sometimes weaker (Worms, Hilborn *et al.*, 2009).

In its overview of international fisheries, the Commonwealth Fisheries Programme has taken a special interest in the situation of inshore and subsistence fishers. More studies are needed, but it seems that amongst the first to suffer as fish stocks decline are the poor coastal communities of the developing world (Bostock and Walmsley; Hanich and Tsamenyi, this volume). They have no voice, receive little development return from industrial fisheries or trade agreements, suffer heavily as a result of IUU and have few alternatives to turn to when their inshore seas become depleted by fishing fleets from afar. If poverty and the MDGs are to be addressed, then the coastal communities that depend upon marine fisheries for their livelihoods must have some measure of long-term sustainable returns.

In their chapter, Crothers and Wilen agree that the continued use of much of the oceans as a common, free good must stop, and they argue that it should be replaced by an extensive rights-based approach to fisheries management, which is proven as successful. These and other authors underscore the fact that effective rights (or allocated shares in fisheries) offer a sustainable and empirically proven means permanently to alter fishing behaviour away from a 'race to fish' to one which is more economically rational. In effect, this model emphasises 'privatisation' rather than public ownership of fishery resources.

Under rights-based regimes the value of economic rents[2] is capitalised into the value of the fishing right, and can be realised upon exchange or transfer. This may provide the incentive to optimise production efficiency while protecting the asset as a useful by-product. However, the rights-based approach remains controversial largely due to perceptions of inequity, and political and economic factors.[3] In practice, a combination of rights-based approaches within a strong framework of regulation on fishing gear and access will need to be developed according to local needs and conditions.

Rights-based approaches can be consistent with ecosystem approaches, and findings, highlighted in this submission by Lang and Trebilco, and by Watts and Bourne, have reported successes from the introduction of Marine Protected Areas (MPAs). A few MPAs in Belize and elsewhere preceded the 2002 World Summit on Sustainable Development, but this was the UN meeting that gave the MPA process a big fillip. However the Fisheries Programme has noted that there remain problems of funding and personnel with some MPAs, and that their continuance is often fragile. It would be sensible for the Commonwealth to share best practice here, so that MPAs become more viable, and important components of effective management systems.

In terms of unlocking the inherent wealth of fisheries to boost development, the World Bank and FAO have reported that global marine capture fisheries are underperforming by at least $50 billion per year (World Bank and FAO, 2009). This is equivalent to more than half the value of the global seafood trade, and over the last three decades the cumulative economic loss to the global economy has been estimated at $3 trillion. However, they have argued that improved governance can recover a substantial proportion of these losses, while at the same time putting capture fisheries on a sustainable footing.

The work of the Commonwealth Fisheries Programme (Watts and Bourne, this volume) has confirmed that there is indeed considerable scope for new investment in aspects of marine fisheries, and relevant capacity building, which the Commonwealth Secretariat and donors have already supported to some extent (see boxed section below). But the programme has found a disconnect between rents charged by governments for commercial fishing – especially tuna rents – and investment in subsistence fishing, the nurturing of fishing cooperatives, and hygienic landing sites and refrigerated transport. Commonwealth member states should be encouraged to evaluate the economic and social contribution of their subsistence fishers in the informal economy, especially for food security in coastal communities, and invest more in their future. The Commonwealth Fund for Technical Cooperation could provide expert advice.

## Marine Fisheries and the Commonwealth

It is worth pausing to recognise the strength and potential of the Commonwealth. Its role has been widely acknowledged in the 20th century campaign against racism in southern Africa. Its lengthy work, going back to an Expert Group report in the 1980s and a meeting of Finance Ministers in Trinidad and Tobago in the early 1990s, led to debt write-off for the poorest countries (the HIPC programme); and its ministerial mission at the end of the Uruguay Round helped to break the diplomatic logjam and create a World Trade Organization.

Albeit not always under a Commonwealth banner, Commonwealth member states have done crucial work on fisheries over many years, for example in the negotiation of the UN Convention on the Law of the Sea, 1982 and, working with NGOs, in the High Seas Task Force, 2003.[4] The Commonwealth has the expertise and the economic, social and environmental interest amongst its member states to play a bigger and more coordinated and focused role in the battle to safeguard marine fisheries, and to put them on a sustainable use basis. While the Commonwealth Secretariat has been able to provide technical assistance to member states in a variety of ways (see below), the Commonwealth's resources and influence have not yet been mobilised on the wider question of sustainability, economic returns and equitability.

A new focus on fisheries is important for the whole spectrum of Commonwealth countries and will provide a unifying theme. Whether industrial, artisanal or subsistence, sustainable fisheries are key to sound economies, food security, poverty eradication and environmental stability. The Commonwealth Fisheries Programme, and the report to Commonwealth leaders which is embodied in the present book, provide evidence of the issues and pointers to the way forward.

## What is the Commonwealth Secretariat doing?

The Commonwealth Secretariat's activities encompass all fisheries sectors: from marine capture fisheries (both ocean and coastal fishing) through to inland capture fisheries and aquaculture (encompassing mariculture). In recent times, a significant proportion of the Commonwealth Secretariat's support has been provided to the Pacific region – which is largely dominated by marine capture fisheries. Elsewhere, the Commonwealth Secretariat support has largely focussed on other fisheries sectors – including support to aquaculture policy and strategy development (e.g. Brunei Darussalam and Mozambique) as well as institutional and capacity building (e.g. Namibia).

Broadly speaking, the Commonwealth Secretariat support aims at four main outcomes:

- Promoting trade in fish and fish products for developing countries;
- Sustainable fisheries business development;
- Sustainable fisheries policies and strategies at the national and regional level;
- Enhanced skills and knowledge of personnel of local fisheries-related agencies and institutions.

### Promoting trade in fish and fish products for developing countries

The international trade in fish and fish products is of particular importance to many Commonwealth countries. But there is a need for trading partners to give more consideration to the potentially adverse impacts on the growth and trading opportunities for Commonwealth countries arising from the adoption of particular trade policies and/or the creation of unnecessary barriers to trade in fish and fish products. This is particularly true with respect to rules of origin (ROO).

The Commonwealth Secretariat has provided support to the ACP region to raise awareness on linkages between the WTO fisheries subsidies negotiations, the ACP-EU EPA negotiations, and their bilateral fisheries partnership agreements. One meeting fostered dialogue between trade and fisheries officials and resulted in the preparation of a joint ACP-wide strategy on fisheries negotiations – in both WTO and ACP fora. At the same time, the Commonwealth Secretariat, in collaboration with the relevant regional Secretariats, facilitated meetings for trade experts from CARICOM, Indian Ocean Commission and the Pacific. These resulted in a joint WTO negotiating proposal on fisheries subsidies which was submitted to the relevant WTO negotiating body; since then two other follow-up proposals have been presented with the support of Commonwealth-funded consultants based at the Pacific Islands Forum office, and the ACP office in Geneva.

Since 2007 the Secretariat has contributed to dialogues which have led to a common strategy for countries in the Africa-Caribbean-Pacific regions, in their discussion with European Development Ministers. It is now providing analysis and supporting consensus-building to assist trade officials in linking trade and climate change policies. The purpose is to formulate strategies, in the WTO and in regional fora, which will ensure that fisheries trade can contribute to sustainable growth in Commonwealth member states.

The Commonwealth Secretariat will continue to help member countries to enable them to pursue their fisheries development objectives by enhancing their participation in international trade negotiations. It will go on calling for greater coherence between trade policies and fisheries policies, and to press for more explicit commitments in support of the fisheries sector in both national-led strategies and donor-supported development programmes in Commonwealth countries.

## Assisting fisheries business development

The fishery resources of several Commonwealth countries offer significant potential for economic development and enhanced food security, particularly through value addition and improved resource pricing.

The Commonwealth Secretariat has in recent times provided support to developing and strengthening the entrepreneurial and business management skills of small and medium sized enterprises (SMEs) through the ILO-based programme 'Start and Improve Your Fisheries Business'.

Assistance will continue to be required to strengthen the capacity of Commonwealth member states in a number of areas, including development of fisheries business enterprise and business management, fish handling, quality control, and food safety, especially in SMEs.

**Support to the design and implementation of sustainable fisheries policies and strategies at national and regional level**

Effective marine fisheries management in Commonwealth countries will largely hinge on the design and implementation of appropriate fisheries policies and fisheries management plans and strategies. Many Commonwealth countries continue to be hampered by a lack of skilled personnel and weak legal frameworks and regulatory instruments. These challenges are further compounded by major data deficiencies, including on fish stocks.

The Commonwealth Secretariat has supported advocacy and technical assistance for the development of regional plans that harmonise national policies pertaining to sustainable management of marine fisheries resources (particularly migratory and straddling fish stocks). For example, the Commonwealth Secretariat recently provided technical assistance to the Pacific region in developing a regional coastal fisheries policy and strategic management plan based on the ecosystems approach (Apia Policy).

A recent Economic Paper, *Fairer Fishing?*, published by the Commonwealth Secretariat, highlights the practical challenges facing many developing countries, including Commonwealth countries, in implementing aspects of international fisheries treaties and other fisheries-related obligations (see also Barnes, Appendix 1, this volume). These include explicit conditions for the adoption of more resource-intensive vessel monitoring and control systems, as well as the implementation of more stringent reporting and verification requirements regarding catch certification and validation to deter Illegal Unreported and Unregulated fishing.

The Commonwealth Secretariat will continue to support policy and other relevant research to ensure that the potential implications of some of these provisions, in terms of the additional resource and capacity requirements they demand, are better understood in Commonwealth countries. Where necessary, appropriate support measures can be put in place. The Commonwealth Secretariat will also continue to offer technical assistance to facilitate the development of national and, where appropriate, regional-level fisheries policies and strategies. It will provide member states with a more secure basis for the management of fisheries and renewable seabed resources through better demarcation of maritime boundaries.

**Enhancing the skills and knowledge in fisheries-related agencies and organisations**

Fisheries personnel in many Commonwealth member countries lack the relevant skills and knowledge, particularly in the areas of fisheries conservation, management and development.

From 2005 to 2008, the Commonwealth Secretariat, in collaboration with the Government of Iceland, supported a programme of development assistance to strengthen the capacity of national fisheries agencies and institutions in Pacific island countries to manage their coastal fisheries sustainably. This included support to the design and delivery of short courses focused on fisheries policy and planning as well as fish stock assessments and statistics.

The Commonwealth Secretariat will continue to support such capacity building.

Source: T. Williams and Commonwealth Secretariat colleagues (personal communications).

## Building a Commonwealth Mandate

Commonwealth Heads of Government have often shown the way forward for the international community by making progressive declarations that cut across the usual political blocs and geographical affiliations. In 2009, meeting in Trinidad and Tobago just prior to the Copenhagen summit on the UN Framework Convention on Climate Change, the Commonwealth summit has a special opportunity. Heads can include a powerful commitment for the sustainable management of marine fisheries in the context of the need for food security for their people, following up on the 2007 Vava'u Declaration on Pacific Fisheries Resources,[5] and at the same time address climate change, which may disproportionately affect developing world fisheries, in line with the Lake Victoria Climate Change Action Plan[6] (Potts and Rocke; Sumaila and Cheung, this volume).

Specifically, in Trinidad and Tobago, the Commonwealth Heads of Government might wish to consider a statement that would:

- Call on the world's governments to rescue global fish stocks, thereby contributing to eradication of poverty and maintenance of food security;
- Reduce and eventually eliminate those types of fishing and fishing agreements which are driving down stocks, and instead press for binding international and regional instruments to sustain fish stocks;[7]
- Call for an end to perverse legal, financial and institutional incentives and subsidies that drive overcapacity and overfishing within an already overcrowded industry;
- Promote progressive management schemes attuned to local needs, mindful that schemes based around fishing rights can provide long term incentives to generate wealth from and conserve fisheries;
- Learn from experiences with and promulgate ecosystem based management, no-take and recovery strategies, including gazetting of Marine Protected Areas;
- Fully enforce bans on illegal (IUU) fishing both within Economic Exclusion Zones (EEZ) and on the High Seas, prosecute illegal fishers and closely monitor capture fisheries in order to safeguard future resources;
- Recognise the social, economic, environmental and nutritional value of the informal fisheries economy, and support small-scale and artisanal fishing;
- Support consumer movements and the commercial food industry in moves to source marine fish from verifiable, sustainable sources and price fish to reflect scarcity.[8]

At the same time, such a statement would benefit from the establishment of a Commonwealth Ministerial Task Force to examine and recommend practical policies to support fisheries restoration in EEZ and territorial waters of the Commonwealth. The Task Force would be supported by the Commonwealth Secretariat, which has a fine record in bringing together governments, and the Commonwealth Foundation, which has an established role in mobilising civil society, scientists and academics. The Commonwealth Fisheries Programme, comprising the Commonwealth Foundation, the Commonwealth Policy Studies Unit (part of London University) and the Commonwealth Human Ecology Council (a non-governmental body), together with its sponsoring governments, is an example of such cooperation.

In several Commonwealth states (such as Kenya, Namibia, New Zealand) progressive and determined approaches to find solutions to overfishing have proven themselves to be valuable to national economies. It is proposed that the Ministerial Task Force convene a Commonwealth Fisheries Conference, to include all major stakeholders, in order to share experience of these and other successes, and discuss appropriate ways forward.

## Putting Policy into Practice

A Commonwealth Ministerial Task Force and Fisheries Conference would stimulate practical ways forward, which might include:

- Recommending proven, wealth-generating and ecosystem based management strategies utilising guaranteed and properly policed fishing rights and executed well within maximum sustainable yields;
- The promotion of satellite surveillance of IUU fishing and collaboration in policing;
- A review of policies in those 14 Commonwealth states that currently offer flags of convenience;
- Improved political cooperation, and strengthening the conservation and protection obligations of regional fisheries management organisations (RFMOs);
- Support for expanding MPA coverage and strengthening their financial and managerial viability; and
- Practical support for artisanal fishers.[9]

Commonwealth agencies will need not only to continue to collaborate with each other, but also to work closely with the Committee of Fisheries (COFI) of FAO and the OECD's Committee for Fisheries, which influences the deepwater fleets of developed states. There is a strong case for a meeting at official level that could examine how Commonwealth agencies, working with member states, scientists and fishery interests, could cooperate with COFI and FAO, particularly in defeating IUU fishing and the misuse of flags of convenience, and promoting a convention for the sustainability of near-shore stocks and the rights of subsistence fishers. The team on the Caribbean study tour, conducted for the Commonwealth Fisheries Programme, advocated the creation of a network of academic, private sector and non-governmental experts in the Commonwealth to assist the work of governments in combating depletion of fish stocks, promoting sustainability strategies and the long-term viability of coastal communities. Commonwealth nations can also play a positive role in the current WTO negotiations regarding the disciplining of harmful subsidies that lead to overcapacity and overfishing (Von Moltke, this volume).

There are of course several specific measures that the Commonwealth, member governments, and relevant interests can either take themselves or stimulate others to advance. One important measure would be to obtain political and practical traction by strengthening the information on stocks, catches, depletion and recovery evidence, industrial and artisanal vessels and numbers of persons involved in fisheries. This is particularly urgent in developing states and where regionally harmonised approaches are needed for improved management of transboundary or shared resources (Haughton; Hanich and Tsamenyi; Tom'tavala, this volume).

The proposed Commonwealth Ministerial Task Force could stimulate speedy initiatives. At present there are gaps in knowledge, and not all that is known is being shared. In the case of commercial tuna stocks, for example, the Commonwealth Fisheries Programme found plans for new processing plants for which there are unlikely to be sufficient stocks; this is a valuable migratory species, where fish can swim from Southeast Asia to the East African coast, and where a number of governments and trawler fleets have a great interest in sustainability.

Capacity building is required at many levels – in government, amongst marine scientists and economists, in the private sector and civil society, and in fishing communities. A commitment by Commonwealth donor nations will be important here, as will the allocation of scholarships and exchange opportunities, especially to persons from small island developing states.[10] Community development assistance to strengthen the organisation of fishers would be helpful in many countries; they are often too weak and fragmented to enter national policy discourse, or improve their collective marketing and incomes. Capacity building for good governance can also stem the risk of corruption in fisheries management, and encourage development returns to coastal communities.

Commonwealth leaders should consider establishing a special, time-limited Fisheries Fund to promote an effective Commonwealth contribution to this global challenge. The Commonwealth Fisheries Programme has been concerned to stop the depletion of marine fish stocks, to promote the sustainable management of fisheries, and the long-term viability of coastal communities. Heads of Government may wish to vary such priorities, but if the Commonwealth is to provide leadership in the world, and assistance for its member states, it is essential that there is a committed budget. Partnering for a more equitable and sustainable future, the theme which Heads will advocate in Port of Spain, requires tangible investment.

There are precedents for this:

- In the 1980s the Commonwealth provided a special budget to support what was then a non-member, Mozambique, which was suffering grievously as a result of its opposition to apartheid.
- In 1989, at the Kuala Lumpur Commonwealth summit, the leaders created a special environmental budget, attached to the Commonwealth Fund for Technical Cooperation, in the run-up to the Earth Summit in Rio de Janeiro.
- In the early 1990s there was a special budget to support the human rights work of the Commonwealth Secretariat.

These were all voluntary budgets, but often subscribed to by many member states. The same format could apply to marine fisheries.

## Conclusion

In the case of marine fisheries we have a pre-eminently Commonwealth issue, deriving from the association's unique maritime heritage, trading patterns and geography. There are differences of emphasis between Commonwealth governments, and between government, fishing industry and consumer interests inside each country. But the Commonwealth is an association with family ties and habits of cooperation, and long practice in defining and resolving differences. Achieving the sustainability of marine fisheries is one sphere where the Commonwealth can provide global leadership.

It is the conclusion of this report that the Commonwealth can and should exert itself to protect and enhance its heritage of marine capture fisheries and the ecosystems, fish populations and species on which the industry depends, addressing issues of governance, rights, legal frameworks, institutions, social equity, economic development, climate change and environmental sustainability. If the Commonwealth can achieve this, each of its two billion citizens, and their future generations, will reap the benefit.

## References

World Bank and FAO (2009). *The Sunken Billions: The Economic Justification for Fisheries Reform.* World Bank and FAO.

Worm, B., Hilborn, R., Baum, J.K., Branch, T.A., Collie, J.S., Costello, C., Fogarty, M.J., Fulton, E.A., Hutchings, J.A., Jennings, S., Jensen, O.P., Lotze, H.K., Mace, P.M., McClanahan, T.R., Minto, C., Palumbi, S.R., Parma, A.M., Ricard, D., Rosenberg, A.A., Watson, R., Zeler, D. (2009). Rebuilding global fisheries. *Science* 325, pp. 578–585.

## Notes

1 Marine capture fishery refers to all kinds of harvesting of naturally occurring living resources in marine environments.

2 Examples of successful fisheries – i.e. ones that generate wealth, protect the environment and are truly sustainable – deal with *economic rents*. We know that rents are the driving force of overfishing. We also know that the failure to deal with them has led to the currently declining industry which costs tax payers some $50–80 billion per year in foregone wealth and in cost-reducing subsidies. This is having a massive impact on developing country economies that we are only now beginning to understand.

3 The debate about privatisation versus public ownership of marine fisheries is in full swing. Whilst chapters in this book have tended to favour the rights-based approach, this may in part be due to an emphasis on Commonwealth developing countries that are relatively small, geographically isolated

and/or where societies have a tradition of community management. In other situations, e.g. Canada, there is evidence that 'armchair' fishermen and investors can harm communities and livelihoods – the opposite of what was intended. For an excellent analysis see *http://www.ecotrust.ca/fisheries/cautionarytale*

4 *http://www.high-seas.org/*

5 *http://www.forumsec.org/_resources/article/files/2007%20Forum%20Communique,%20Vava'u%20-%20Final%20Version.pdf*

6 The Lake Victoria Commonwealth Climate Change Action Plan 2007 recognised that 'most environmental problems transcend national boundaries, and therefore require solutions that are mutually reinforcing at global, regional, national and community levels; and ... require active participation by all'.

7 Building on the principles established by the international community in the Code of Conduct for Responsible Fisheries, Agenda 21 and the Johannesburg Declaration on Sustainable Development (see Haughton, this volume).

8 A wide range of such schemes is described by Tindall, this volume.

9 Of the kind undertaken by the Instituto Nacional de Desenvolvimento de Pesca de Pequena Escala in Mozambique, the National Institute for the Development of Small-scale Fisheries.

10 E.g. funding for the Caribbean Fisheries Training and Development Institute in Trinidad and Tobago to bring students from neighbouring islands, as was once provided by Japanese aid.

# Editors

**Richard Bourne** is Associate Fellow at the Commonwealth Policy Studies Unit (CPSU ) and Senior Research Fellow at the Institute of Commonwealth Studies, London University. Earlier a journalist, he has been involved in Commonwealth activities since 1982, when he became Deputy Director at the then Commonwealth Institute, Kensington. He was first Director of the Commonwealth Human Rights Initiative (1990), prior to its transfer to New Delhi, and first Head of the CPSU (1999) until retiring in 2005. He has written several books; in 2008 he edited *Shridath Ramphal: the Commonwealth and the World* (Hansib) and authored *Lula of Brazil: the story so far* (University of California Press and Zed Books). He chairs the board of *The Round Table*, the Commonwealth journal of international affairs.

**Mark Collins** is Director of the Commonwealth Foundation, an intergovernmental body established in 1965 to strengthen civil society organisations and professional networks across the Commonwealth. Previously he served as Director for Biodiversity Conventions in the UN Environment Programme, Nairobi, Kenya, and as Director of the UNEP World Conservation Monitoring Centre in Cambridge, UK. Mark studied Natural Sciences at Wadham College, Oxford, gained a PhD in ecology and entomology from Imperial College, London, and a Master's in Business Administration from the Open University. He is a Commonwealth Scholarships Commissioner and sits on the boards of the Galapagos Conservation Trust, *The Round Table* and Friends of the Commonwealth and has authored or edited 12 books and 135 publications, mostly on environmental or civil society themes.

# Contributors

**Michael Arbuckle** is a Senior Fisheries Specialist with the World Bank in Washington DC. He has held senior level position in government and the private sector in New Zealand and more recently has provided specialist advice in fisheries management for the United Nations Food and Agriculture Organisation and the World Bank on a range of developing country fisheries.

**Richard Barnes** is Senior Lecturer, Law School, University of Hull, UK and managing editor of *International Fisheries Law and Policy*. He has research interests in all areas of International Law but especially the Law of the Sea and international maritime law and is engaged in a number of interdisciplinary projects.

**Per Erik Bergh** is Executive Director, Southern Africa Office, Nordenfjeldske Development Service, a Norwegian consultancy company that works in the area of natural resource management, policy and planning. He has a background in the Royal Norwegian Navy and Coastguard followed by more than 12 years of international work in the field of fisheries and marine development and collaboration.

**Tim Bostock** is Fisheries Programme Coordinator and Policy Adviser at the Department for International Development (DFID) and Department for the Environment, Food and Rural Affairs (Defra), London. He is a specialist in renewable natural resources management and research with over 30 years experience with inland and coastal fisheries. Tim established and ran the Support Unit for International Fisheries and Aquatic Resources (SIFAR), for six years at FAO, Rome, and previously worked on long-term projects in Latin America, South and Southeast Asia, and Africa.

**William Cheung** is Lecturer in Marine Ecosystem Services at the University of East Anglia, Norwich U.K. His research focuses on assessing impacts of fishing and climate change on marine ecosystems and their services, extinction vulnerability assessment, ecosystem modelling and evaluating trade-offs between ecological and socio-economic objectives in managing exploited marine ecosystems.

**G.T. (Stan) Crothers** is Deputy Chief Executive, Ministry of Fisheries, New Zealand. He has worked in the Ministry for over 30 years and has been involved in State Sector Reform, establishing the New Zealand EEZ and the implementation of New Zealand's Quota Management System. He has provided fisheries management advice to Chile, Argentina, Kenya, Australia, Pacific island states and the FAO.

**Sandy Davies** is Technical Director, Southern Africa Office, Nordenfjeldske Development Service, a Norwegian consultancy company that works in the area of natural resource management, policy and planning.

**Quentin Hanich** is Senior Research Fellow, Australian Centre for Ocean Resources and Security (ANCORS), University of Wollongong, Australia.

**Milton O. Haughton** is Deputy Executive Director, Caribbean Regional Fisheries Mechanism, Belize.

**Marta Lang** is a New Zealand lawyer. She teaches international environmental policy at the Oxford School of Geography and the Environment, is a Fellow at ANCORS, Australia, and directs the conservation consultancy Our Nature. Our Nature acts as a strategic bridge between funders and communities across the Asia-Pacific. Marta is engaged on Commonwealth Fisheries Programme work in the Pacific Islands and South Asia. Her academic research interests include fisheries management, marine conservation, coastal management and community-led conservation and development.

**Tom McClurg** is Director, Toroa Strategy Limited. He has 18 years experience in senior government and private sector executive positions within the New Zealand fisheries sector including 10 years with Maori organisations (the Treaty of Waitangi Fisheries Commission and Aotearoa Fisheries Limited).

**Arthur C. Potts** is a Fisheries and Marine Resource Management Specialist and the substantive Administrator, Division of Education, Youth Affairs and Sport, Tobago House of Assembly (THA). He is currently seconded to the University of Trinidad and Tobago (UTT) as Senior Research Associate, in the Marine Sciences Programme at the Institute of Marine Affairs.

**Judy Rocke** is Manager, Technical Advisory Services at the Institute of Marine Affairs (IMA), Trinidad and Tobago.

**Rashid Sumaila** is Associate Professor and Director of the Fisheries Economics Research Unit at the University of British Columbia Fisheries Centre. He specialises in bio-economics, marine ecosystem valuation and the analysis of global issues such as fisheries subsidies, IUU (illegal, unreported and unregulated) fishing and the economics of high and deep seas fisheries. He has experience working in fisheries and natural resource projects in Norway, Canada and the North Atlantic region, Namibia and the Southern African region, Ghana and the West African region and Hong Kong and the South China Sea.

**Charlotte Tindall** is a Consultant within the social and economic development department of MRAG Ltd, a firm of Marine and Fisheries Consultants dedicated to promoting sustainable utilisation of natural resources. She specialises in socio-economics, aquatic resource management and communications, and has recently been working on fisheries certification. This has involved assessing options for integrating social and economic issues into voluntary market standards; assessing impacts of aquaculture certification on small-scale producers and undertaking a feasibility study for Fairtrade on options to engage with fisheries products. She has worked on fisheries and environmental projects in Uganda, Mali, Benin, Senegal, Tanzania, Mauritius, Jordan and the Caribbean.

**Yoli Tom'tavala** is a Lecturer at the School of Law, University of the South Pacific. His teaching and research interests include customary law, environmental law, law of the seas and international human rights law.

**Rowan Trebilco** is a PhD student in marine conservation at Simon Fraser University, Vancouver. Rowan worked for the Australian Antarctic Division before taking up a Rhodes scholarship to complete his MSc(dist) in Biodiversity, Conservation and Management at Oxford University.

**Martin Tsamenyi** is Professor of Law and Director of the Australian Centre for Ocean Resources and Security (ANCORS), University of Wollongong, Australia.

**Anja von Moltke** is with the Division of Technology, Industry and Economics, UNEP, Geneva, Switzerland.

**Suzannah F. Walmsley** is a Fisheries Consultant with MRAG Ltd, a leading natural resources management consultancy. She specialises in fisheries socio-economics and policy, small-scale fisheries in developing countries, food security, fisheries trade and co-management. She has experience working in Africa, Latin America, Europe and Asia.

**Nicholas Watts** is Principal Lecturer, Department of Applied Social Sciences, London Metropolitan University and Education Adviser to the Commonwealth Human Ecology Council. He specialises in the social science analysis of environmental issues, sustainability and environmental ethics.

**James Wilen** is Professor of Agricultural and Resource Economics and Director of the Center for Natural Resource Policy Analysis at the University of California, Davis. His research is in the fields of natural resource economics and environmental economics, with particular interests in bioeconomic modelling, fisheries policy analysis, spatial models of resource use, and institutional solutions to problems of the commons.

# Index